HYDRAULIC MOTORS.

TRANSLATED FROM THE

FRENCH COURS DE MÉCANIQUE APPLIQUÉE.

PAR

M. BRESSE

Professeur de Mécanique à L'École des Ponts et Chaussées.

BY

F. A. MAHAN,

LIEUTENANT U. S. CORPS OF ENGINEERS.

REVISED BY

D. H. MAHAN, LL.D.,

PROFESSOR OF CIVIL ENGINEERING, &C., UNITED STATES MILITARY ACADEMY.

NEW YORK:

JOHN WILEY & SON, 2 CLINTON HALL, ASTOR PLACE.

1869.

THE NEW YORK PRINTING COMPANY,
81, 83, *and* 85 *Centre Street*,
NEW YORK.

PREFACE.

The eminent position of M. Bresse in the scientific world, and in the French Corps of Civil Engineers, is my best apology for attempting to supply a want, felt by the students of civil engineering in our country, of some standard work on Hydraulic Motors, by furnishing a translation of the chapter on this subject contained in the second volume of M. Bresse's lectures on Applied Mechanics, delivered to the pupils of the School of Civil Engineers (*École des Ponts et Chaussées*) at Paris.

In making the translation, I have retained the French units of weights and measures in the numerical examples given, as the majority of our students are conversant with them.

F. A. MAHAN.

Willett's Point, N. Y., *July*, 1869.

CONTENTS.

§ IV.—OF A FEW MACHINES FOR RAISING WATER.

APPENDIX.

HYDRAULIC MOTORS,

AND SOME

MACHINES FOR RAISING WATER.

---

§ 1.—Preliminary Ideas on Hydraulic Motors.

1. *Definitions; theorem of the transmission of work in machines.*—The term *machine* is applied to any body or collection of bodies intended to receive at some of their points certain forces, and to exert, at other points, forces which generally differ from the first in their intensity, direction, and the velocity of their points of application.

The *dynamic effect* of a machine is the total work, generally negative or resisting, which it receives from external bodies subject to its action. It happens that the dynamic effect is sometimes positive work: for example, when we let down a load by a rope passed over a pulley, the weight of the load produces a motive work on the system of the rope and pulley.

Let us suppose, to make this clear, that the dynamic effect is a resisting work. Independently of this, the machine is affected by some others which are employed to overcome friction, the resistance of the air, &c. The resistances which give rise to these negative works have received the name of *secondary resistances;* whilst the dynamic effect is due to what are called *principal resistances.*

The general theorem of mechanics, in virtue of which a relation is established between the increase of living force of a material system and the work of the forces, is applicable to a machine as to every assemblage of bodies. To express it analytically, let us suppose the dynamic effect to be taken negative, and let us call

T_m, the sum of the work of the motive forces which have acted on the machine during a certain interval of time;

T_e, the dynamic effect during the same time;

T_f, the corresponding value of the work of the secondary resistances;

v and v_0, the velocities of a material point, whose mass is m, making a part of the machine, at the beginning and end of the time in question;

H and H_0, the corresponding distances of the centre of gravity of the mechanism below a horizontal plane;

Σ, a sum extended to all the masses m.

From the theorem above mentioned there obtains

$$\tfrac{1}{2}\Sigma m v^2 - \tfrac{1}{2}\Sigma m v_0^2 = T_m - T_e - T_f + (H - H_0)\Sigma m g.$$

In a machine moving regularly, each of the velocities v increases from zero, the value corresponding to a state of rest, to a certain maximum which it never exceeds; the first member of the equation has, then, necessarily, a superior limit, below which it will be found, or which it will, at the farthest, reach, whatever be the interval of time to which the initial and final values v_0 and v are referred. The same holds with the term $(H - H_0)\Sigma m g$, when the machine moves without changing its place, as its centre periodically occupies the same positions. On the contrary, the terms T_m, T_e, T_f increase indefinitely with the time, if the motion of the machine is prolonged, because new quantities of work are being continually added to those already accumulated. These terms will at length greatly surpass the others, so the equation, therefore, should, after an unlimited interval of time, reduce to

$$T_m = T_e + T_f.$$

This is what would really take place, without supposing the time unlimited, if the beginning and end of this interval corresponded to a state of rest of the machine, and if, at the same time, $H = H_o$. We can then say that, as a general rule, the motive work is equal to the resisting work; but as this last includes, besides the dynamic effect for which the machine is established, the work of the secondary resistances, we see that the action of the motive power is not all usefully employed, since a portion goes to overcoming the work of T_f.

It is evident that the ratio $\frac{T_f}{T_m}$ measures the proportional loss; the ratio $\frac{T_e}{T_m}$, or $1 - \frac{T_f}{T_m}$, gives, on the contrary, a clear idea of the portion of the effective work. This last ratio is what is called *the delivery* of the machine; it is evidently always less than unity, since in the best-arranged machines T_f still preserves a certain value. The skill of the constructor is shown in bringing this as near unity as possible.

2. *Analogous ideas applied to a waterfall.*—A waterfall may be considered as a current of water flowing through two sections C B, E F (Fig. 1), at no great distance apart, but with a noticeable difference of level, H, between the surface slope at C and E, and which may be assumed to yield a constant volume of water for each unit of time. The material liquid system thus comprised between the sections C B, E F, at each instant may be regarded as a machine that is continually renewed, the molecules which flow out through E F

being replaced by those which enter through C B. The motive work in this case will be that of the weight combined with that of the pressures on the external boundaries of the system; the dynamic effect will be the work of the resistances against the fall of water caused by any apparatus whatever, a water-wheel

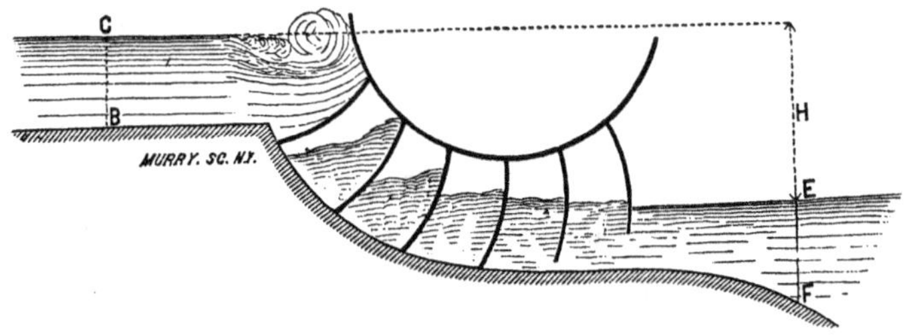

FIG. 1.

for example, exposed to its action. In its turn the water-wheel, considered as a machine, will receive from the fall a motive work sensibly equal to the dynamic effect that we have just spoken of,* and will change only a portion of it into useful work, which will be its dynamic effect proper. But in what is to follow we shall limit ourselves to studying the dynamic effect of the fall, and not that of the wheel.

Although the motion of the liquid cannot always be strictly the same, because the wheel does not always maintain exactly the same position, still it can be so regarded without material error; for after an interval of time θ, generally very short, occupied by a float or paddle in taking the place of the one that preceded it, everything returns to the same condition as at the beginning of the interval. Supposing the motion of the wheel regular and the paddles to be uniformly distributed, there is such a frequent periodicity in the state of the system that it

* We say *sensibly*, for the equality between the mutual actions of the water and wheel do not involve that of corresponding work. This equality is only strict in supposing the friction of the liquid against the solid sides of the wheel zero, which friction is in reality very slight.

almost amounts to a permanency. We will now apply Bernouilli's theorem to any molecule whatever, having a mass m, which, departing from the section C B with the velocity V_0, reaches E F with a velocity V. The entire *head* is H, if we allow the parallelism of the threads in the extreme sections, for the pressure then varies according to the hydrostatic law in each of the two surfaces C B and E F, so that the points C and E can be considered as piezometric levels for the initial and final positions of the mass m. Let $-t_e$ and $-t_f$ be the respective work referred to the unit of mass, and considered as resistances which m has encountered in its course between C B and E F, in consequence of the action of the wheel and of viscosity. Then from the general theorem of living forces we have the equation—

$$H - \frac{1}{g}(t_e + t_f) - \frac{V^2 - V_0^2}{2g} = 0;$$

whence,

$$m\, t_e = m\, g\left(H + \frac{V_0^2}{2g} - \frac{V^2}{2g}\right) - m\, t_f.$$

Consequently we see that the weight $m\,g$ of each molecule which passes from C B to E F gives rise to a dynamic effect $m\,t_e$, the value of which is expressed in the second member of the equation. The sum of the weights of the molecules m included in the entire weight P, which the current expends in a second, will produce a dynamic effect equal to a sum Σ of analogous expressions extended to all these masses; considering V_0 and V as constant velocities in the respective sections C B and E F, this summation will give—

$$\Sigma\, m\, t_e = P\left(H + \frac{V_0^2}{2g} - \frac{V^2}{2g}\right) - \Sigma\, m\, t_f \quad . \; . \; . \; . \quad (1)$$

Moreover, in each second a new weight P is supplied by the current; there is then produced a new dynamic effect $\Sigma\, m\, t_e$, which thus represents the mean dynamic effect in each second.

The quantity $P\left(H + \frac{V_0^2}{2g} - \frac{V^2}{2g}\right)$ reduces to P H, in the case in which V_0 and V are sensibly equal to zero, which happens in measuring the difference of level between the basins from which the water starts and that into which it flows, when the water is nearly at a stand-still; we then call the product P H the *effective delivery of the fall*. The ratio $\frac{\Sigma\, m\, t_e}{P\, H}$ is the *productive force;* $\Sigma\, m\, t_f$ is the work lost. Dividing the last equation by P or $\Sigma\, m\, g$, and supposing $V = 0$, $V_0 = 0$, there obtains,

$$\frac{\Sigma\, m\, t_e}{g\, \Sigma\, m} = H - \frac{\Sigma\, m\, t_f}{g\, \Sigma\, m} \quad . \; . \; . \; . \quad (2)$$

H in this expression may be regarded as the total head of water; $\frac{\Sigma\, m\, t_e}{g\, \Sigma\, m}$ the *productive head*, that is, the height which, multiplied by the weight P expended, would give the dynamic effect per second; $\frac{\Sigma\, m\, t_f}{g\, \Sigma\, m}$ the *head lost*, to be subtracted from the total head when the productive head is required. We see that $\frac{\Sigma\, m\, t_f}{g\, \Sigma\, m}$ is the mean loss of head experienced by the molecules in their passage from C B to E F; for this expression represents exactly the mean work of viscosity on a molecule, referred to the unit of mass and divided by g.

3. *General remarks on the means of securing a satisfactory delivery from a head of water which moves an hydraulie motor.*—In order to obtain a good satisfactory delivery, we must seek to diminish as much as possible the term $\Sigma\, m\, t_f$, or the mean loss of head $\frac{\Sigma\, m\, t_f}{g\, \Sigma\, m}$. A few of the causes that produce this loss will now be pointed out, and the manner in which they may be diminished.

Firstly, if the water enters the wheel, and can in consequence act on it, it is because it possesses a certain relative velocity w; now, in the majority of cases this relative velocity gives rise to a violent agitation of the liquid and to vibratory motions, from which a loss of head is experienced equal to $\frac{w^2}{2g}$, like to what has been observed to obtain in the collision of solid bodies; hence $\frac{w^2}{2g}$ would be a portion of the head lost.* It is, then, generally a matter of importance to make the water enter with as small a relative velocity as possible. However, that is not necessary when w has its direction tangent to the sides that it comes in contact with, and when the particular arrangement of the apparatus allows the water to continue its relative motion in the wheel without there being any shock of the threads on the solid sides or on the liquid already introduced, since then we have no longer to fear the violent disturbance that we have just spoken of.

When the water on leaving the wheel is received into a race of invariable level, in which it loses its absolute velocity of

* Let us suppose that the water that has once entered the wheel passes at once to relative rest: the destruction of the velocity w being then attributable only to the resisting work of the molecular actions, this work for a fluid molecule having the mass m would be $\frac{1}{2} m w^2$, a quantity which, when referred to the unit of mass and divided by g, would give the loss of head $\frac{w^2}{2g}$, which is the same for all the molecules. But the supposition of the instant production of relative rest is not, strictly speaking, true; weight, for example, can combine sometimes with molecular actions to bring about this result at the end of a sensible time. Consequently the expression $\frac{w^2}{2g}$ must be considered less as the exact value of the quantity whose value we here wish to find, than as a superior limit which we approach more or less nearly, according to the particular case considered.

departure v', we readily see that this velocity v' must be reduced as much as possible. In fact, the water that leaves the wheel carries with it, in each second, a living force, $\frac{P\ v'^2}{2\ g}$, which might have been taken up by a resisting work of the hydraulic motor, and have thus increased by its amount the dynamic effect T_e; whereas this living force will only serve to produce a disturbance and eddies in the interior race, and will enter the term $\Sigma\ m\ t_f$. There are cases, however, in which we are obliged to give v' a value more or less considerable; we shall see presently, by a few examples, how it is sometimes possible to diminish this unfavorable condition.

Ordinarily, the considerations above mentioned are understood when we say that the water must enter without shock, and leave without velocity. We can also add, that it is well not to deliver it too rapidly through channels of too little breadth, as this would involve losses of head to be included in the expression $\frac{\Sigma\ m\ t_f}{g\ \Sigma\ m}$.

We shall now proceed to examine the most widely known hydraulic motors, keeping principally in view the best means of making use of the head in each case, and showing the manner of calculating the dynamic effect that can be realized with the means adopted.

Water-wheels are divided into two great classes—those having a vertical and those having a horizontal axis. The varieties included in these two classes will form the subject of the following paragraphs.

UNDERSHOT WHEELS.

---

§ II. Water-Wheels with a Horizontal Axis.

4. *Undershot wheel with plane buckets or floats moving in a confined race.*—These wheels are ordinarily constructed of wood. Upon a polygonal arbor A (Fig. 2) a socket C, of cast iron,

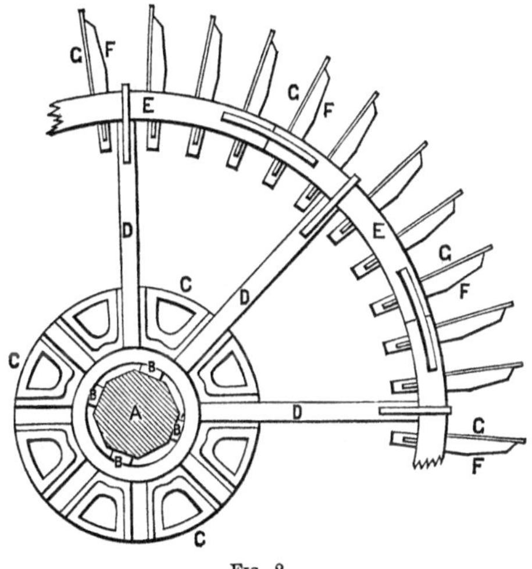

Fig. 2.

is fastened by means of wooden wedges *b*. Arms D are set in grooves cast in the socket, and are fastened to it by bolts; these arms serve to support a ring E E, the segments of which are fastened to each other and to the arms by iron bands. In the ring are set the projecting pieces F F...., of wood, placed at equal distances apart, and intended to support the floats G G. . . . , which are boards varying from $0^m.02$ to $0^m.03$ in thickness, situated in planes passing through the axis of the wheel and occupying its entire breadth. A single set of the foregoing parts would not be sufficient to give a good support to the floats. In wheels of little breadth in the direction of the axis two parallel sets will suffice; if the wheels are broad, three or more may be requisite.

The number of arms increases with the diameter of the wheel. In the more ordinary kinds, of 3 to 5 metres in diameter, each socket carries six arms. The floats may be about $0^m.35$ to $0^m.40$ apart, and have a little greater depth in the direction of the radius, say $0^m.60$ to $0^m.70$.

From this brief description of the wheel, let us now see how we can calculate the work which it receives from the head of water. The water flows in a very nearly horizontal current through a race B G H F (Fig. 3), of nearly the same breadth

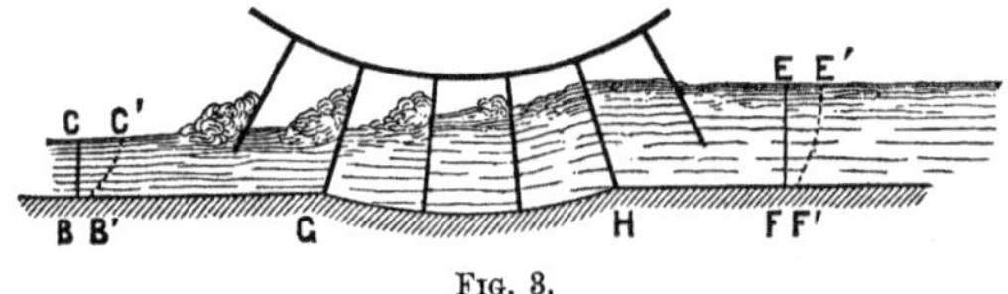

Fig. 3.

as the wheel, a portion G H of the bottom being hollowed out, in a direction perpendicular to the axis, to a cylindrical shape, and allowing but a slight play to the floats. The liquid molecules have, when passing C B, a velocity v, but shortly after they are confined in the intervals limited by two consecutive floats and the race. They entered these spaces with a mean relative velocity equal to the difference between the horizontal velocity v, and the velocity v' of the middle of the immersed portion of the floats, the direction of which last velocity is also very nearly horizontal. There result from this relative velocity a shock and disturbance which gradually subside, while the floats are traversing over the circular portion of the canal; so that if this circular portion is sufficiently long, and if there be not too much play between the floats and the canal, the water that leaves the wheel will have a velocity sensibly equal to v'. The action brought to bear by the wheel on the water is the cause of the change in this velocity from v to v', which gives us the means, as we shall presently see, of calculating the total intensity of this action.

For this purpose let us apply to the liquid system included between the cross sections C B, E F, in which the threads are supposed parallel, the theorem of quantities of motion projected on a horizontal axis. Represent by

b the constant breadth of the wheel and canal;

h, h' the depths $\overline{\text{C B}}$, $\overline{\text{E F}}$, of the extreme sections which are supposed to be rectangular;

F the total force exerted by the wheel on the water, or inversely, in a horizontal direction;

P the expenditure of the current, expressed in pounds, per second;

Π the weight of a cubic metre of the water;

θ Θ the short interval of time during which C B E F passes to C′ B′ E′ F′.

The liquid system C B E F, here under consideration, is analogous to the one treated in *Note A* (see Appendix), in which a change in the surface level takes place; and the manner of determining the gain in the quantity of motion during a short time θ, and calculating the corresponding impulses, during the same time, produced by gravity and the pressures on the exterior surface of the liquid system, are alike in both cases.

Employing the foregoing notation, we obtain

1st. For the mean gain in the projected quantity of motion, $\frac{P\theta}{g}(v'-v)$;

2d. For the impulses of the weight and pressures together, also taken in horizontal projection, $\frac{1}{2}\Pi b \theta (h^2 - h'^2)$; to these impulses is to be added that produced by F, or $-F\theta$, to have the sum of the projected impulses.

We will then have

$$\frac{P\theta}{g}(v'-v) = \frac{1}{2}\Pi b \theta (h^2 - h'^2) - F\theta,$$

whence we obtain

$$F = \frac{P}{g}(v-v') - \frac{1}{2}\Pi b \theta (h'^2 - h^2).$$

The forces of which F is the resultant in horizontal projection are exerted in a contrary direction by the water on the wheel, at points whose vertical motion is nearly null, and whose horizontal velocity is approximately v'. The wheel will then receive from these forces, in each unit of time, a work $F v'$, which represents the dynamic effect T_e to within a slight error. So that

$$T_e = F v' = \frac{P}{g} v'(v-v') - \frac{1}{2}\Pi b v' (h'^2 - h^2).$$

Moreover we have

$$P = \Pi b h v = \Pi b h' v';$$

whence

$$T_e = \frac{P}{g} v'(v-v') - \frac{1}{2}P\left(h' - \frac{h^2}{h'}\right) = \frac{P}{g} v'(v-v') - \frac{1}{2}Ph$$
$$- \frac{1}{2}Ph\left(\frac{h'}{h} - \frac{h}{h'}\right);$$

or finally, observing that $\frac{h}{h'} = \frac{v'}{v}$

$$T_e = \frac{P}{g} v'(v-v') - \frac{1}{2}Ph\left(\frac{v}{v'} - \frac{v'}{v}\right) \quad \ldots \ldots (1)$$

In order that this formula should be tolerably exact, the depths h and h' must be quite small, without which the floats would make an appreciable angle with the vertical at the moment they leave the water; the velocity of the points at which are applied the forces, whose resultant is F, could no longer be considered horizontal, as heretofore, and a resistance due to the emersion of the floats would be produced, on account of the liquid uselessly raised by them. The water must also be confined a sufficiently long time to assume the velocity v'.

We can consider in formula (1), v and h as fixed data, and seek the most suitable value of the velocity v' of the floats, to make T_e a maximum. For this purpose, if we place $\frac{v'}{v} = x$, $T_e = A\,P\,\frac{v^2}{2\,g}$, formula (1) becomes

$$A = 2\,x\,(1 - x) - \frac{g\,h}{v^2}\left(\frac{1}{x} - x\right),$$

and we must choose x so as to make A a maximum. This value will be obtained by taking the value which reduces the first differential co-efficient $\frac{d\,A}{d\,x}$ to zero, which gives the equation

$$1 - 2\,x + \frac{g\,h}{2\,v^2}\left(\frac{1}{x^2} + 1\right) = 0.$$

The value of x deduced depends on $\frac{g\,h}{v^2}$; we find approximately

For $\frac{g\,h}{v^2} = 0.00$	$x = 0.500$	$A = 0.500$
0.05	0.553	0.431
0.10	0.595	0.373

Generally we neglect the term $\frac{g\,h}{v^2}\left(\frac{1}{x} - x\right)$; then we have

$x = 0.50$, $A = 0.50$; and recollecting that the maximum value of A is very much changed, even for small values of $\frac{gh}{v^2}$. Besides, experiment does not show that there is any use in taking x greater than 0.50, as we have just found; it would rather show a ratio of v' to v of about 0.4, which obtains probably because of the motion of the wheel being too swift to allow the liquid to pass completely from the velocity v to v', during the time that it remains between the floats: a portion of the water passes without producing its entire dynamic effect, and the formula (1) ceases to conform to fact.

Smeaton, an English engineer, made some experiments, in 1759, on a small wheel, $0^m.609$ in diameter, having plane floats. This wheel was enclosed in a race having a flat bottom, which is a defect, because the intervals included between the floats and the race are never completely closed. The weight P varied from $0^k.86$ to 2.84 kilogrammes. In each experiment the work transmitted to the wheel was determined by the raising of a weight attached to a rope which was wound around the axle. The most suitable value of $x = \frac{v'}{v}$ was thus found to vary between 0.34 and 0.52, the mean being 0.43. The number A, comprised between 0.29 and 0.35, had for a mean value about $\frac{1}{3}$.

We can then take definitively the ratio $\frac{v'}{v} = 0.4$, as found by experiment. The expression for A becomes then

$$A = 0.48 - 2.1 \frac{gh}{v^2},$$

which, for $\frac{gh}{v^2} = 0.05$ and $\frac{gh}{v^2} = 0.10$, gives the numbers

$$A = 0.375 \text{ and } A = 0.27,$$

very nearly those found by Smeaton.

A few remarks remain to be made, to which it would be well to pay attention in practice.

1st. It is well, as far as possible, to have the depth h say from $0^m.15$ to $0^m.20$. Too small a thickness of the stratum of water which impinges on the wheel would give a relatively appreciable importance to the unavoidable play between the wheel and the race, a play which results in pure loss of the motive water. Too great a thickness has also its inconveniences: for from the relation $\frac{h'}{h} = \frac{v}{v'}$, it follows that for $\frac{v}{v'} = 0.4$ and $h = 0^m.20$, we will then have

$$h' = \frac{0.65}{0.04} = 0^m.50\,;$$

the floats will then be immersed $0^m.50$, adopting the thickness of $0^m.20$, and were they more so, they would meet with considerable resistance in leaving the water, as we have already said. It is necessary then that h be neither too large nor too small: the limits of $0^m.15$ to $0^m.20$ are recommended by M. Belanger.

2d. We should avoid as much as possible losses of head during the passage of the water from the basin up-stream to its arrival at the section C B near the wheel, losses which result in a diminution of effective delivery (No. 2). In order to give it the shape of a thin layer, from $0^m.15$ to $0^m.20$ in thickness, the water is made to flow under a sluice through a rectangular orifice: care should be taken to avoid abrupt changes of direction between the sides of this orifice and the interior of the dam, in order to avoid a contraction followed by a sudden change of direction of the threads, as in cylindrical orifices. The sluice should be inclined (as in Fig. 4), in order to leave but a small distance between the orifice and the wheel, which will diminish the loss of head produced by the friction of the water on the portion of the race M C, through which the water reaches the wheel.

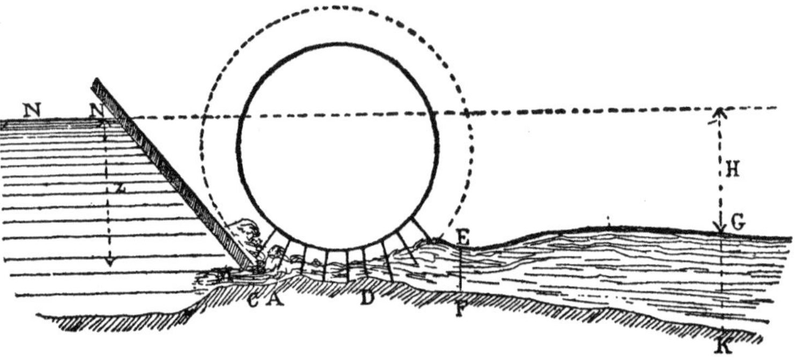

FIG. 4.

3d. The water leaves the wheel at E F with a velocity $v' = 0.4\ v$, in the shape of a horizontal band of parallel threads. If, in order to flow into the tail race through a section G K, where its velocity will be sensibly zero, it had to undergo no loss of head, there would be between E F and G K a negative head, the absolute value of which would be $\frac{v'^2}{2g}$, or $0.16\ \frac{v^2}{2g}$; that is, the point G would be at a height $0.16\ \frac{v^2}{2g}$ above E, since the piezometric levels at E F and G H may be confounded with those of the points E and G. It is not possible so to arrange every part that all loss of head shall be suppressed between E F and G K; but these losses are much diminished by means of a plan first recommended by M. Belanger. The bottom of the race, beyond the circular portion A D, presents a slight slope, for a distance D I = 1 or 2 metres; thence it connects with the tail race by a line I K, having an inclination from $0^m.07$ of a metre to $0^m.10$ for each metre in length; the side-walls are prolonged for the same distance, either keeping their planes parallel, or very gradually spreading outwards, but never exceeding 3 or 4 degrees. The point D is placed at a height $h' + \frac{2}{3}$, $0.16 \frac{v^2}{2g}$ (*), or $2.5\ h + 0.11 \frac{v^2}{2g}$ below the level of the water in the tail race. From this the following effects take place: the water overcomes the difference of level between E and G, or the height $0.11 \frac{v^2}{2g}$, in virtue of its velocity $0.4\ v$, either by a surface counter-slope, or by a sudden change of level with a counter-slope, so that the head lost reduces to $0.05 \frac{v^2}{2g}$.

In many wheels this precaution has been neglected, and the level of the tail race has been placed at the same height as the point E, and sometimes even below it. We will now show that a loss of effective delivery is thus produced. For this purpose let us see what must, with the above-described arrangement, be the position of the race relative to the pond, and the expression for the effective delivery. In view of simplifying this investigation, we will admit that the lines M A, D F, slightly inclined, are a portion of the same horizontal. In the hypothesis of no loss of head up to B C, the velocity v would be due to the height z of the level N in the pond, above the highest thread of the fluid vein thrown on the wheel; but, on

* The co-efficient $\frac{2}{3}$ is simply assumed: in replacing it by unity the water would no longer be able to attain the level of the point G, since this would require a loss of head which would be null in the interval between E F and G K; consequently the water in the lower race would probably drown the floats and impede their motion. It is for the purpose of avoiding such an inconvenience that the number in question is taken less than unity; the value assumed gives a disposable head expressed by $\frac{1}{3} \frac{v'^2}{2g}$, or $0.05 \frac{v^2}{2g}$, to counterbalance the loss of head of the liquid molecules after escaping from the wheel, and to secure for the wheel their free discharge.

account of the losses of head, we give z a co-efficient of reduction, which we will take (for want of precise data) equal to 0.95; that is, we will write

$$\frac{v^2}{2g} = 0.95\, z.$$

whence

$$z = \frac{1}{0.95} \cdot \frac{v^2}{2g}.$$

The distance from the bottom M A to the level N will then be $h + \frac{1}{0.95} \frac{v^2}{2g}$; this can also be expressed in another way, thus:

$$\text{H} + 2.5\, h + \frac{2}{3}, 0.16 \frac{v^2}{2g},$$

by calling H the total head, or the difference in height between G and N. Hence

$$h + \frac{1}{0.95} \frac{v^2}{2g} = \text{H} + 2.5\, h + \frac{2}{3}, 0.16 \frac{v^2}{2g};$$

and consequently

$$\frac{v^2}{2g} = 1.057 \left(\text{H} + \frac{3}{2} h\right),$$

a relation which gives v, for any given head, when the value of h has been determined. From this we can deduce z and $h + z$, which is sufficient to determine the position of the race. The dynamic effect T_e, from what we have just seen, will be $\text{A P} \frac{v^2}{2g}$, or, making $\frac{v'}{v} = 0.4$,

$$\text{T}_e = \left(0.48 - 2.1 \frac{g\, h}{v^2}\right) \text{P} \frac{v^2}{2g}; \quad \ldots \quad (2)$$

substituting for $\frac{v^2}{2g}$ its value, this relation becomes $\text{T}_e = \text{P} \left[0.48 \times 1.057 \left(\text{H} + \frac{3}{2} h\right) - 2.1 \cdot \frac{1}{2} h\right] = \text{P}\ (0.507\ \text{H} - 0.289\, h)$

The effective delivery $\frac{T_e}{P\,H}$ will then be expressed in round numbers by

$$\frac{T_e}{P\,H} = 0.50 - 0.3\,\frac{h}{H};$$

for $h = 0^m.20$ and H included between 1 metre and 2 metres, it would vary from 0.44 to 0.47.

Now let us suppose that, without changing the head H, we wish to place the level of the tail race below E, or, at most, on the same level; it is plain that we will have to raise the bottom of the race. Then v will diminish, and h will have to increase in order that the expenditure P may remain the same; for these two reasons T_e will diminish, as the above formula (2) shows. For example, supposing that the tail race is at the height of E, the equation that determines $\frac{v^2}{2\,g}$ will become

$$h + \frac{1}{0.95}\,\frac{v^2}{2\,g} = H + 2.5\,h,$$

whence we derive successively, regard being had to equation (2),

$$\frac{v^2}{2\,g} = 0.95\left(H + \frac{3}{2}\,h\right)$$

$$T_e = P\left[0.48 \times 0.95\left(H + \frac{3}{2}\,h\right) - 1.05\,h\right] = P\,(0.456\ H -$$

$$0.375\ h),$$

and in round numbers

$$\frac{T_e}{P\,H} = 0.45 - 0.375\,\frac{h}{H}.$$

The data $h = 0^m.20$ and $H = 1$ metre would give a effective delivery of 0.38 instead of 0.44; with $H = 2$ metres, we would obtain an effective delivery of 0.41, whilst we had found 0.47.

There would be a much more marked falling off, if we supposed the bottom of the tail race on the same level as the bottom of the portion preceding it, as was the manner of constructing the race formerly.

The arrangement of the channel through which the water flows off, which we have mentioned as by M. Belanger, can be advantageously employed in all systems of hydraulic wheels from which the water flows with a sensible velocity, in the form of a horizontal current, with parallel threads. The rising of the surface of this current taking place beyond the wheel, this latter will experience the same action from the water, if everything is similarly arranged from the head race to the outlet of the wheel. With the canal in question, the level rises, instead of remaining the same or falling; we then obtain the same action on the wheel with a less head, and consequently we can have a greater effect the head remaining the same.

However, we see by formula (3) that the effective delivery of these wheels never reaches 0.50, in spite of all possible precautions; this system is, then, not comparatively as good as those which we are now going to take up.

5. *Wheels arranged according to Poncelet's method.*—The principal cause of loss of work in the undershot wheel with plane floats is the sudden change from the velocity v to v', twice and a half less, which necessarily produces in the liquid a violent disturbance. From this disturbance proceed great inner distortions and a negative work produced by viscosity, all of which diminishes the dynamic effect. The water also possesses a great velocity of exit, which is at best only partially turned to account. General Poncelet proposed to avoid these inconveniences, continuing, however, to preserve to the wheel its special character, which is rapid motion; that is, he has

endeavored to fulfil for the undershot wheel the two general conditions of a good hydraulic motor, viz.: the entrance of the water without shock, and its exit without velocity. To this end he has contrived the following arrangements:

The bottom of the head race, which is sensibly horizontal, is joined without break to the flume, the profile of which is composed of a right line of $\frac{1}{10}$, followed by a curve.

The right line forms a slope near the wheel, and its prolongation would be tangent to the outer circumference of this latter; it ends at the point at which the water begins to enter the wheel. The curved portion is composed of a special curve, to the shape of which we will return presently, and which stops at the point at which it meets the exterior circumference. Finally, the floats are set in a cylindrical portion of the race, having a development a little greater than the interval beween two consecutive floats, and terminated by an abrupt depression; this depression has its summit at the mean level of the water down-stream; its object is to facilitate the discharge of the water from the wheel.

The water enters the race under a sluice inclined at an angle of from 30 to 45 degrees with the vertical; the sides of the orifice are rounded off, so as to avoid the loss of head analogous to that in cylindrical orifices.

The floats are set between two rings or shrouds, which prevent the water from escaping at the sides; the interior space between the rings is somewhat greater than the breadth of the orifice opened by the sluice. The floats are curved; they intersect the outer circumference of the crown at an angle of about 30 degrees, and are normal to the inner circumference; beyond this, their curvature is a matter of indifference. There are ordinarily 36 for wheels of from 3 to 4 metres in diameter, and 48 for those from 6 to 7 metres.

The exact theory of this wheel it is almost impossible to explain in the present state of science. It is simplified, first, by considering the mass of water that enters between two consecutive floats as a simple material point, which, during its relative motion in the wheel, would experience no friction. We suppose, moreover, that the absolute velocity of this point is in the direction of the horizontal that touches the wheel at its lowest point, and that the floats are so put on as to be tangent to the exterior circumference. Now, let v be the absolute velocity of the water on striking the wheel, and u the velocity on this circumference; the water possesses relatively to the wheel, at the moment of entering the floats, a horizontal velocity $v - u$, in virtue of which it takes up a motion towards the interior of the basin formed by two consecutive floats If we liken, during this very short relative motion, the motion of the floats to a uniform motion of translation along the horizontal, the apparent forces will reduce to zero, so that the small liquid mass that we have spoken of will ascend along the floats to a height $\frac{(v - u)^2}{2g}$, on account of its initial relative velocity, which is gradually destroyed by the action of gravity. Then this mass descends, and again takes up the same relative velocity, $v - u$, when on the point of leaving the float; but this relative velocity will be in a direction contrary to the preceding, and, consequently, also in a direction contrary to the velocity u of the floats. The absolute velocity of the water on leaving will then be equal to the difference between u and $v - u$, or $2u - v$; we see that it will be zero if we have $u = \frac{1}{2} v$, that is, provided that the velocity at the outer circumference of the wheel is half that of the water in the supply channel.

We could thus realize the two principal conditions for a good wheel. But, as M. Poncelet has observed, such favorable circumstances are found by no means in practice.

The water cannot enter the wheel tangent to its circumference, as we have here supposed. In fact, let us call ds the length of an element of this circumference immersed in the current, b the breadth of the wheel, β the angle formed by ds and the relative velocity, w, of the water referred to the wheel; there will enter during a unit of time, through the surface bds, a prismatic volume of liquid having for a right section $bds.$ sin β, and a length w; in other words, a volume $bds.w$ sin β, a quantity that reduces to zero, for $\beta=0$. Hence w must intersect the circumference at a certain angle which cannot be zero; besides, we must make it as small as possible, in order that, at the point of exit, the relative velocity of the liquid and the velocity of the floats may be sensibly opposite, and give a resultant zero; on the other hand, it must not be so small as to make the entrance of the water difficult or impossible. It is in order to reconcile these two contradictory conditions that the angle β has been fixed at 30 degrees, which is also that made by the floats with the exterior circumference, since the threads must enter in the direction tangent to the floats. But then the absolute velocity of the water is no longer zero on leaving the wheel, for its two components are no longer following the same right line, but make an angle of $180°-30°$ or $150°$. Now, supposing $v-u=u$, the resultant v' would have for a value $2u \cos 75°$, or $v \cos 75°$, or finally, $0.259v$; this resultant lies furthermore in the direction of the bisecting line of the angle between u and $v-u$; that is to say, it is almost vertical, and consequently it is impossible to turn it to account by means of a counter-slope; its effect is destroyed in producing a disturbance in the lower portion of the canal, whence there results a negative work equal to P $\frac{v'^2}{2g}$, P being the expenditure of the head. It is then a loss of head expressed by $\frac{v'^2}{2g}$, or $0.067\frac{v^2}{2g}$.

As a cause of loss we may still mention the friction of the water against the race and against the floats. Another very serious objection to the above-mentioned theory is that the liquid molecules do not move as though they were entirely isolated; when one of them, having reached the height $\frac{(v-u)^2}{2g}$ between the floats, is in its descent, another has just entered, and it is not clear that no sensible disturbance in the motion of the molecules follows.

On account of all these reasons, experiment indicates only an effective mean work of 0.60 for these wheels. Nevertheless, the improvement is very great on the old undershot wheels with plane floats, whose effective work was scarcely 0.25 or 0.30, and could never reach the limit 0.50. As to the most favorable ratio between the velocities w and v, experiment gives it 0.55 instead of $\frac{1}{2}$.

It has been observed that the straight portion of the race was followed by a curve; the following condition determines its form. The relative velocity of the water $v - u$ at its entrance, equal to that which exists at its exit, is also the same as the circumference velocity u; then the direction of the absolute velocity, resulting from these two velocities, is on the line bisecting the angle formed by these two, and since the angle between the tangent to the exterior circumference and the relative velocity is 30 degrees, that between the same tangent and the absolute velocity will be 15 degrees. All the threads, then, must make an angle of 15 degrees with the circumference. To deduce from this the shape of the bottom, let us assume that every normal to the bottom is at the same time normal to all the threads that it intersects. Let A then (Fig. 5) be the point of entrance of a thread, the absolute velocity A v of which makes an angle of 15 degrees with the tangent A u; the perpendicular B A B′ to Av is a normal to the bottom of the race. At the same time, if we draw the radius A O, the angle O A B will be equal to uAv, as their sides are perpendicular; hence O A B′ = 15 degrees. The perpendicular O B′ let fall from the centre O to the prolongation of B A, has then a constant value equal to A O sin 15°. Consequently, all the normals to the bottom of the race are tangent to the same circle, having a radius O B′; the curve C B D is then the involute of which this circle is the evolute. It is terminated at one end by the circumference OA, at the point D, and at the other at a point C, such that the normal C E, taken as far as the circumference O A, may have a length equal to the thickness of the fluid stratum. This thickness, moreover, varies with the head; it must be, according to experiment, from $0^{m}.20$ to $0^{m}.30$ for heads below $1^{m}.50$, and may be diminished to about $0^{m}.10$ for heads of more than 2 metres.

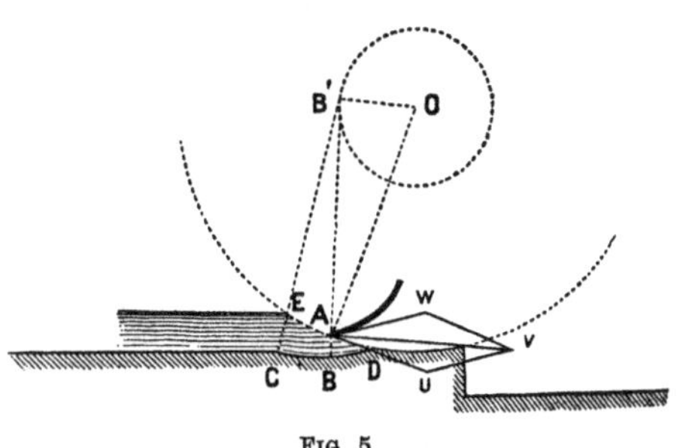

FIG. 5.

The water rises in the float to a height $\frac{(v-u)^2}{2g} = \frac{1}{4}\frac{v^2}{2g}$, which differs little from $\frac{1}{4}$ H, calling H the height of the head.

The distance between the two circumferences that limit the float should be at least $\frac{1}{4}$ H; in order more certainly to avoid the possibility of the water still possessing any relative velocity on reaching the extremity of the floats, which would give rise to a spirting of the water to the interior of the wheel, M. Poncelet has advised increasing this advance to $\frac{1}{3}$ H.

6. *Paddle wheels in an unconfined current.*—These wheels are placed in a current whose section has a breadth much greater than that of the wheel; frequently they are supported by boats, and are called hanging wheels. It being impossible to calculate theoretically the dynamic effect of the current on these wheels, we will be satisfied with the following general ideas.

The horizontal force F which the wheel exerts on the liquid being constant, its impulse in the unit of time has numerically the same value as its intensity. If, then, owing to this impulse, a mass m of water passes, in a second, from the velocity v of the current to the velocity u of the wheel, we shall have

$$F = m\,(v - u),$$

and consequently the work produced on the wheel in the same time, sensibly equal to the dynamic effect of the current, will be

$$F\,u = m\,u\,(v - u).$$

Moreover, General Poncelet supposes that the mass m must be proportional to v, and, furthermore, it is quite natural to admit that it is proportional to the area S of the immersed portion of the floats. He then places, calling B a constant co-efficient and Π the weight of a cubic yard of water,

$$m = B\,S\,v\,\frac{\Pi}{g}$$

whence we have

$$F\,u = B\,\Pi\,S\frac{v\,u\,(v-u)}{g}.$$

This formula has been found quite true by experiment, in taking $B = 0.8$.

When v is given, the maximum state of $F\,u$ corresponds to $u = v - u$, since the sum of these two factors is constant: from this we deduce $u = \frac{1}{2}\,v$, as we found for the two wheels previously discussed. Experiment indicates the ratio $\frac{u}{v} = 0.4$ as being the most suitable; this ratio only changes very slightly the theoretical maximum, the value of which is

$$0.4\,\Pi\,S\,v\,\frac{v^2}{2\,g}.$$

The depth of the floats must be from $\frac{1}{5}$ to $\frac{1}{4}$ the length of the radius. Flanges placed on the edge, on the side which receives the shock of the water, will increase the mutual action. The diameter is ordinarily from 4 to 5 metres, and the floats are 12 in number.

BREAST WHEELS.

7. *Wheels set in a circular race, called breast wheels.*—These wheels resemble very closely, in their construction, the undershot wheels noticed in No. 4; but one essential difference, which can produce a great change in their effective delivery, consists in the method of introducing the water; this no longer enters the wheel at its lower portion, but at a slight depth only below the axle—that is to say, on the side. The questions now to be successively examined will be, 1st, the most suitable velocity of the wheel; 2d, the method of introducing the water; 3d, the situation of the race as regards the upper end and tail race; 4th, the manner of determining the dynamic effect. Finally, several practical ideas will be put forth on the subject of this wheel.

(*a*) *To determine the velocity of the wheel.*—Let Q be the volume that the head expends in a second;

b the breadth of the wheel, which is the same as that of the sluice;

h the depth of the immersed portion of the floats directly beneath the axle;

c the thickness of the floats;

C their distance apart, measured between their axes along the exterior circumference;

R the radius of this circumference;

u the velocity of any one of its points.

The water comprised between any two consecutive floats, directly beneath the axle, will have for its depth h, for its mean breadth $C\left(1-\frac{h}{2R}\right)-c$, and b for its thickness parallel to the axis of the wheel; its volume is then $hbC\left(1-\frac{h}{2R}-\frac{c}{C}\right)$, and, supposing that it expends during one second a number $\frac{u}{C}$ of these volumes, we shall have

$$Q = hbu\left(1-\frac{h}{2R}-\frac{c}{C}\right).$$

In practice $\frac{h}{2R}$ and $\frac{c}{C}$ are always small fractions, the sum of which rarely exceeds 0.10; we would then have, except a slight error,

$$Q = 0.9\,hbu.$$

Q being given, we can satisfy this relation by assuming u and h arbitrarily, and then determining b. In this case, we would take h from $0^m.15$ to $0^m.25$, $u = 1^m.30$, and finally, $b = \frac{10\,Q}{9\,h\,u}$; the expenditure $\frac{Q}{f}$ per metre in breadth, expressed by $0.9\,h\,u$, would then vary from 175 litres to about 270 litres.

The values just given for h and u may be satisfactorily accounted for as follows:

The calculation of the losses of head sustained by the water, in its passage from the head race to the tail race, shows, as we shall see further on, that all these losses increase with the velocity of the wheel. We should naturally be inclined to make this velocity very small, in order to increase the effective delivery. But several conditions show that it is not well to allow the wheel to move too slowly. In fact, we see that if u were very small, the product $hb = \frac{10}{9}\frac{Q}{u}$ would be large, and one of the two dimensions b or h would have to be quite large. Now a great breadth b would give us a heavy wheel, expensive to put up, losing a great deal of work by the friction of the axle; a great value of h would produce inconveniences already mentioned in speaking of the undershot wheel (No. 4). Finally, it is well to have the wheel perform, up to a certain point, the functions of a fly-wheel for the machinery that it sets in motion, which is another reason for allowing it to retain a certain velocity. The value $u = 1^m.30$ has been given by experiment. As regards the depth h, it is well, on account of the unavoidable play between the wheel and the race, that it should not descend much below $0^m.15$, in order not to lose too great a proportion of water.

But it sometimes happens that, in taking $u = 1^m.30$, and h within the limits above mentioned, we arrive at a great value for b, or a value that goes beyond a fixed limit, either on account of local circumstances, or through economy in construction. We are then obliged to increase h or u. It is not well to have h greater than $0^m.45$ or $0^m.50$, and when this limit is reached, we must then begin to increase u.

For example: let $Q = 0^m.50$. Taking $u = 1^m.30$, $h = 0^m.20$, we would have $b = \frac{10}{9}\frac{Q}{hu} = 2^m.14$, a value that in general is quite admissible. But if, by reason of particular circumstances, we could not exceed a breadth of $0^m.71$, about the third of $2^m.14$, we should first increase h, bringing this up to $0^m.50$; we would then find $u = \frac{10\,Q}{9\,b\,h} = 1^m.56$. Or else, as the velocity $1^m.56$ is not yet very great, we would be content to take $h = 0^m.40$, which would give $u = 1^m.96$.

(*b*.) *Method of introducing the water.*—As has already been said (No. 3), in order to diminish as much as possible the loss of work produced by the introduction of the water on the wheel, we must so arrange matters as to have a small relative velocity of the water at its point of entrance, or, when that cannot be obtained, the relative velocity must be tangent to the first element of the floats, and the water must move to the interior of the wheel without shock, merely gliding along the solid sides.

If the wheel be moving slowly—that is, if the velocity at the circumference be about $1^m.30$ per second, we should let the water in by the means indicated in the following figure.

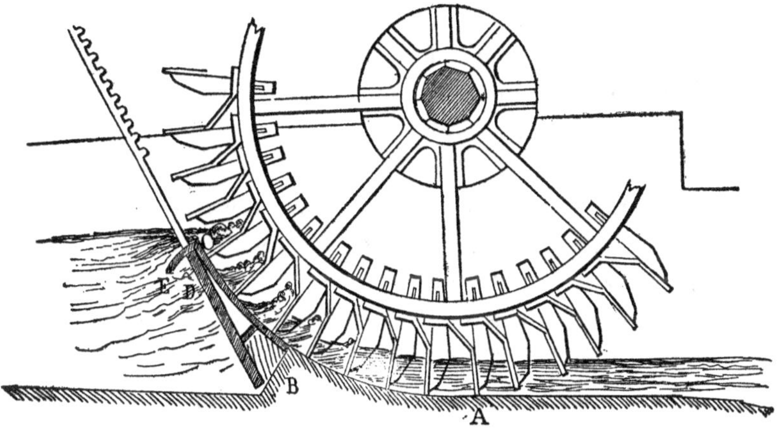

FIG. 6.

The flume A B, constructed of masonry, is prolonged by a piece of cast iron B C, called a *swan's neck*, or guide bucket; A B C is the arc of a circle nearly coincident with the exterior circumference of the wheel, leaving the slightest amount of play. A sluice D, furnished at its upper extremity with a small, rounded metallic appendage E, can slide along the guide bucket while resting on it; a system of two racks connected by cog-wheels allows this sluice to be raised to any convenient point. The metallic appendage E forms the sole of a weir over which the water flows to get to the wheel; this sole must be from $0^m.20$ to $0^m.27$ below the level of the pond.

The sluice being thus very near the wheel, the absolute velocity of the water at its entrance, due only to the slight fall which takes place in the surface of the pond, will consequently be small; and as the wheel itself moves slowly, the relative velocity will be moderate, and the disturbance produced hardly perceptible. We see that, in order to attain this end, we should reduce the head over the sole of the weir, for in the contrary case the velocity of flow would have a greater or less value; we should not, moreover, carry the reduction too far, in order that the loss of water between the wheel and flume may not be too great. The limits $0^m.20$ to $0^m.27$ fulfil this double condition quite well; they correspond very nearly to the limits $0^m.15$ and $0^m.25$ above mentioned for the depth h of the current just beneath the axle; for a weir without lateral contraction can yield about 180 litres per metre of breadth, with a head of $0^m.20$ above the sole, and 280 litres with a head of $0^m.27$.

The rounded metallic appendage E has for its object to diminish the contraction of the sheet of water flowing over the weir, and thus for the same head to yield more water: the head and velocity of flow are then less for a given yield, which diminishes the disturbance of the liquid on entering. The piece E may also serve to direct the vein so as not to intersect the exterior circumference of the wheel at too great an angle; this is as it should be, for, all other things being equal, the relative velocity of the water increases with the angle in question, as the following considerations show, although they more particularly apply to a different case.

Let us now pass to the case of rapidly moving wheels. The velocity u of the wheel at its circumference has been fixed as has been shown; but we still, in order to lessen the loss due to the introduction of the water, dispose of the absolute velocity v of this latter, in intensity and direction. Let, at the point of entrance, M (Fig. 7), M U be the velocity u of the wheel, M V the absolute velocity v of the water, γ the angle formed by these two right lines. The line M W, equal and parallel to U V, will be the velocity w of the water relatively to the wheel. The first point is to make w as small as possible. Now we should have $w = o$, if γ were zero and v equal to u; but it is not possible for the water to enter without any relative velocity, and at an angle zero. One thread for which γ may have the value 0, would be tangent to the exterior circumference of the wheel; but it is evident that the other threads, placed in juxtaposition parallel to this, could not fulfil the same condition, unless the total thickness of these threads taken together were itself zero, which cannot be admitted in the case of a finite expenditure. We have seen that in Poncelet's wheel the angle γ was taken equal to 15 degrees; in the breast wheel it is generally taken to be 30 degrees, in order to somewhat facilitate the introduction of the water. Then, since U is a point determined as well as the direction M, the smallest possible value of U V $= w$ would be the perpendicular let fall from U on M V, or $u \sin \gamma$, or finally $\frac{1}{2} u$ if γ be taken equal to 30°; v would have for a corresponding value $u \cos \gamma$, or 0.866 v, γ being 30°.

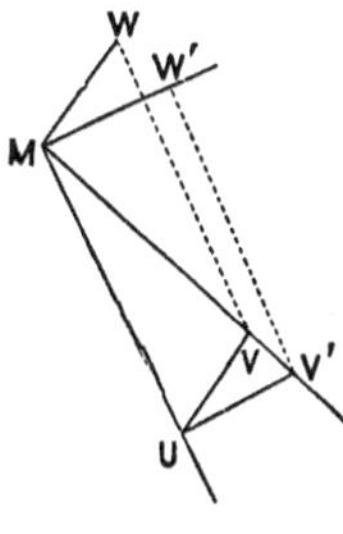

FIG. 7.

Still these values are not those taken for v and w. It is well, in fact, to have the first element of the floats in the direction of w, in order to diminish the disturbance of the water within the wheel. Now, if this first element were perpendicular to M V, it would make an angle of 120° with M U—that is, with the circumference of the wheel, and consequently an angle of 30° with the radius through M. When this radius, by the act of revolving, reaches the vertical, immediately below the axle, the first element of the float will still be inclined at an angle of 30° with the vertical, and it would be still greater at the point of emersion, which would obstruct the motion of the wheel. On this account it is better that the relative velocity M W should pass through the axis of rotation and be perpendicular to M U; the parallelogram of velocities is then represented by the rectangle M U V ′ W ′, in which $\overline{MU} = u$, $\overline{MW'} = w$, $\overline{MV'} = v$. The angle γ still preserving its value of 30°, we have these relations:

$$v = \frac{u}{\cos 30^\circ} = 1.55\ u,$$

$$w = u \text{ tang } 30^\circ = 0.577\ u.$$

The intensity of v and its angle with u being henceforth determined, it remains to be seen how in practice these two conditions can be realized. To obtain them, we will first determine the height $\frac{v^2}{2\ g}$; we will increase it a little, say by one-tenth, in order to compensate approximately for the losses of head sustained by the liquid between the head race and the point of entrance, and $1.1\ \frac{v^2}{2\ g}$ will be the depth of the point of entrance below the level of the head race. The direction of the velocity v will be obtained by drawing through the point of entrance, which we have just found (since it is on the exterior circumference of the wheel, and on a known horizontal line), a line making an angle of 30° with the said circumference. The threads should then be obliged to take this direction by means of a canal of from $0^m.50$ to $0^m.60$ in length, the last element of which should be tangent to v, and into which the water would flow, either by passing under a sluice, or by flowing out without any obstacle, the bottom of this canal being then the sole of the weir.

(c.) *Position of the flume as regards the head race and tail race.*—We have only to repeat here what has already been said in discussing the undershot wheel with plane floats. When the water possesses, on leaving the wheel, a sensible velocity, it is well to place the level of the portion between the floats, which is directly under the axle, below the level of the lower section, by a quantity equal to $\frac{2}{3}\frac{u^2}{2g}$, u being the velocity of the wheel at the circumference. The velocity u, having a value previously determined, as well as the height h of the water between the floats, the situation of the bottom of the flume is then also fixed, at least that part below the axle. On the up-stream side this bottom has a circular profile, with a radius sensibly equal to that of the wheel; it should be joined with the tail race in accordance with M. Belanger's rules, of which we have previously spoken (No. 4).

When the wheel is moving slowly, u being about $1^m.30$, $\frac{2}{3}\frac{u^2}{2g}$ gives a depth a little below $0^m.06$. However great the head may be, the loss or gain of $0^m.06$ is of but small importance, and we would then be able, by economy, to do away with the portion of the race beyond the wheel, or at least to do without a masonry-lined canal with a regular section. The level of the water between the floats would then coincide with that of the tail race. But in the case of large values of u, especially if, at the same time, we have only a slight head, the height $\frac{2}{3}\frac{u^2}{2g}$ may afford a sensible gain that we should not neglect.

(*d.*) *To calculate the dynamic effect of a head of water which sets a breast wheel in motion.*—Calling P the weight of the water expended in a second, H the head measured between the two portions of the canal supposed to be in a state of rest, we know (No. 2) that the dynamic effect T_e of the head, during each second, has for its value the product of the weight P by the height H, diminished by the mean loss of head that the liquid molecules sustain by reason of friction in their passage from one portion of the canal to the other. Moreover, this loss of head is composed of several portions, which we shall now proceed to analyze, using the notation adopted (No. 7, *a*, *b*, *c*).

1st. *Loss of head between the head race and the point at which the water enters the wheel.*—This is reduced by avoiding, by means of rounded outlines, contractions followed by a sudden expansion, and by placing the point of entrance as near the head race as possible. Nevertheless there is always a certain loss: this we may assume at first sight as comprised between $0.05\frac{v^2}{2g}$ and $0.1\frac{v^2}{2g}$. If there were a canal between the wheel and the head race, we should have to take the friction in this canal into consideration, by a calculation analogous to that which will be presently given for the circular flume.

2d. *Loss of head due to the introduction of the water.*—After what has been said (in No. 3), this loss may be valued at $\frac{w^2}{2g}$, or at $\frac{1}{2g}(v^2 + u^2 - 2uv\cos\gamma)$. In the case of slowly moving wheels it is of slight importance, like all the other losses that we consider; in one moving quickly, if $\gamma = 30°$, $v = \frac{u}{\cos \gamma}$, we shall have

$$\frac{w^2}{2g} = \frac{u^2}{2g} \tan^2 \gamma = \frac{1}{3} \frac{u^2}{2g}.$$

When the precaution is taken to introduce the water with a relative velocity tangent to the floats, and a surface is presented along which it can ascend in virtue of this relative velocity, it is probable that the action of gravity assists in overcoming this velocity, and thus so much less effect will be consumed in a violent disturbance of the water. This consideration justifies the use of polygonal floats, such as are shown in (Fig. 6), the arrangement of which was contrived by M. Belanger in 1819. They are composed of three planes making angles of 45 degrees with each other, of which the furthest from the centre is in the direction of a radius, the nearest touches the circumference of the ring, and the third connects the two others. The planes that are fixed to the crown have vacant spaces left between them to facilitate the disengagement of the air. The point of entrance of the water must, moreover, be lower than the centre of the wheel, in order that the water introduced may be received upon an ascending inclined plane.

Data are wanting to estimate the effect produced by the use of this kind of float.

3d. *Loss of head produced by the friction of the water against the circular flume.*—We know that the friction of a current on its bed, per square yard, is expressed by $0.4\ U^2$, U being the mean velocity. Moreover, calling V and W the velocities at the surface and at the bottom, it has been found that these quantities are connected by the approximate relations:

$$U = \frac{1}{2}(V + W),\ U = 0.80\ V;$$

whence

$$U = \frac{4}{3}W.$$

The friction per square yard of bed would then be

$$0.4.\frac{16}{9}W^2 \text{ or } 0.71\ W^2.$$

Now, if L be the length of the circular race, L $(b + 2\,h)$ will be the surface wet, and the entire friction is expressed by 0.71 L $(b + 2\,h)\,u^2$, since the velocity at the bottom is the same as the velocity at the circumference of the wheel. All the points of application of the forces that compose this friction moving with the velocity u, their negative work in the unit of time will be 0.71 L $(b + 2\,h)\,u^3$; we shall obtain the corresponding loss of head by dividing by P, or by 1000 bhu, which gives

$$0.00071\frac{L\,(b + 2\,h)\,u^2}{b\,h}, \text{ or else } 0.014\ \frac{L\,(b + 2\,h)}{b\,h}.\frac{u^2}{2\,g}.$$

4th. *Loss of head from the point just beneath the axle to the tail race.*—If the level of the water between the floats, beneath the axle, exceeds by a height η the level of the portion lower down, the velocity of the water will increase on account of this difference of level. This velocity, sensibly equal to u on leaving the wheel, will become $\sqrt{u^2 + 2\,g\,\eta}$, for the point at which the water begins to enter the lower portion. As we have seen (No. 3), a loss of head $\eta + \frac{u^2}{2\,g}$ corresponds to this velocity.

When we adopt the arrangement of the tail race recommended by M. Belanger (No. 4), η becomes negative and equal to $\frac{2}{3}\frac{u^2}{2\,g}$; the loss then reduces to $\frac{1}{3}\frac{u^2}{2\,g}$.

(*e*) *Some practical data.*—With regard to the number of

arms, the same rule here applies as to undershot wheels (No. 4).

The number of floats is a multiple of the number of arms; their distance apart may be from once and a third to once and a half the head above the top of the weir, in the case of slowly moving wheels. In wheels of rapid motion, it will always be necessary to take this distance apart a little greater at the portion of the circumference intercepted by the stratum of water that falls upon the wheel. It is not well to place the floats too far apart, because some of the threads might fall from quite a considerable height before reaching them; it is also bad to place them too close together, because the water could with difficulty enter the wheel, and a portion would be thrown off.

The depth of the floats in the direction of the radius is but little greater than $0^m.70$. In its normal condition, the interior capacity formed by any two consecutive floats should be very nearly double the volume of water that they contain: we might then take the depth in question equal to $2\ h$, whenever h is not greater than $0^m.35$.

The diameter of the wheel should be at least $3^m.50$; it is seldom greater than 6 or 7 metres. The axis of the arbor is placed a little above the level of the up-stream portion of the canal.

Breast wheels are suitable for falls of from 1 to 2 metres, or even $2^m.50$. Beyond these limits they can still be frequently used to advantage.

When a breast wheel of slow motion is well organized, the dynamic effect of the head may approximate quite near to what is due to the entire head. General Morin, in some experiments on these wheels, found an effective delivery of 0.93. But, since the measure of the expenditure of water is always somewhat involved in uncertainty, and consequently the total effect due to a given head can be but imperfectly ascertained, it will be prudent not to count beforehand on an effective delivery, in practice, greater than 0.80.

8. *Example of calculations for a rapidly moving breast wheel.*—The wheel in question was experimented on by General Morin; it belonged to the foundry at Toulouse.

The water left the head race under a sluice that was raised $0^m.147$ above the sole, which was $1^m.423$ below the above-mentioned level. The orifice being prolonged by a very nearly horizontal race, the velocity v with which the water reaches the wheel will be due to the head at the upper portion of the orifice, except a co-efficient of correction very nearly 1, which we will value at 0.95; hence

$$v = 0.95 \sqrt{2g\,(1^m.423 - 0^m.147)} = 4^m.75.$$

The velocity at the circumference of the wheel was $u = 3^m.06$, and the angle made by u and v, at the point of introduction, was valued at 30°. It follows directly from this, that the loss of head sustained in bringing the water from the head race to the wheel should be $\frac{1}{10}$ of the head $(1^m.423 - 0^m.147)$ or $0^m.13$; we can also calculate that produced by the introduction of the water, which was expressed by (No. 7, *d*)

$$\frac{w^2}{2\,g} = \frac{1}{2\,g}\left(v^2 + u^2 - 2\,u\,v\cos 30°\right) = 0^m.34.$$

The level of the water between the floats, just below the axle, being at the level in the down-stream portion of the canal, we must again count upon a loss equal to $\frac{u^2}{2\,g}$ (No. 7, *d*), or $0^m.48$.

Finally, there is a loss in the circular flume. The depth of the water beneath the axle and the breadth of the wheel being valued respectively at $0^m.20$ and $1^m.55$, and the length of the flume being $2^m.50$, we find for this loss $0.014.\frac{2.50 \times 1.95}{0.20 \times 1.55}\frac{u^2}{2g}$, or $0^m.11$.

All these losses together make up a head of $0^m.13 + 0^m.34 + 0^m.48 + 0^m.11$, or of $1^m.06$. The head being $1^m.72$, we see that the productive force as calculated would be only $\frac{1.72 - 1.06}{1.72}$, that is, 0.38; M. Morin found experimentally 0.41, a number which corresponds to an available head of

$$1^m.72 \times 0.41 = 0^m.705,$$

instead of $0^m.66$, which the preceding calculation gives. This difference, otherwise hardly noticeable, from $0^m.045$, belongs probably to a somewhat greater value of the head lost by the introduction of the water; in fact we have seen that, in certain cases, a portion of the relative velocity could be annulled by the action of gravity, which would diminish by so much the disturbance within the floats, and would give rise to a smaller loss of head.

To increase the effective delivery of this wheel without changing its velocity, the following arrangements might have been made: First, to place the flume at $\frac{2}{3}\frac{u^2}{2g}$, or $0^m.32$ lower than its actual position, taking care to arrange the race as described by M. Belanger (No. 4)—that is, without any sudden variation in section, and with a bottom having a moderate slope, to its junction with the tail race; then to raise the point of entrance of the water so as to reduce the velocity v to $\frac{u}{\cos 30^\circ}$ or $\frac{3^m.06}{0.866} = 3^m.52$. The loss from the wheel to the tail race would then have been reduced to $\frac{1}{3}\frac{u^2}{2g}$ or to $0^m.16$, instead of $0^m.48$; the loss for the entrance of the water would also be reduced to the same value $0^m.16$ instead of $0^m.34$, which would procure a total benefit of $0^m.50$. The other losses remaining sensibly the same, the available head would be $0^m.66 + 0^m.50 = 1^m.16$, and the effective delivery would be raised to about $\frac{1.16}{1.72} = 0.67$.

OVERSHOT WHEELS.

9. *Wheels with buckets, or over-shot wheels.*—These wheels are not, like the preceding, set in a canal. The water is let in at the upper portion; it enters the buckets, which are, as it were, basins formed by two consecutive floats, terminated at the sides by the annular rings, and closed at the bottom or sole by a continuous cylindrical surface concentric with the wheel. The questions that the organization of this kind of motor present are as follows:

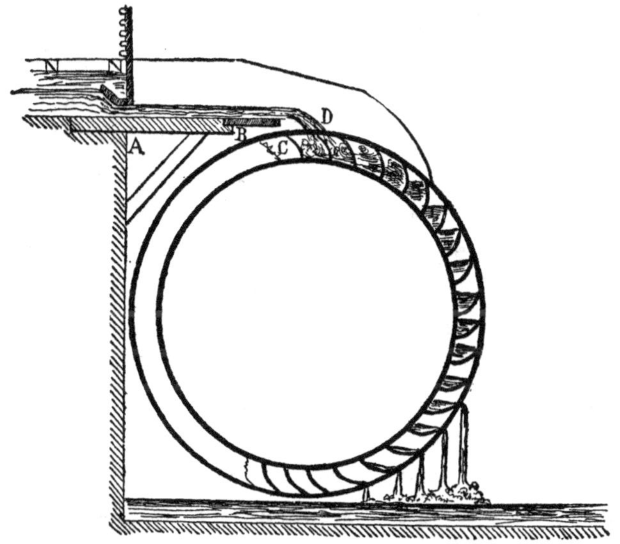

FIG. 8.

(*a*) *Introduction of water into the wheel.*—Two arrangements are employed which are represented hereafter (Figs. 8, 9). In (Fig. 8) the top D of the wheel is placed a little below the level N N of the pond, at $0^{ft}.60$ to $0^{ft}.75$ lower than that level; the water is led to a point C, situated about $1^{ft}.50$ up-stream, and is delivered directly above the axle, by means of a canal A B, or pen trough made of planks, terminated by a very thin metallic plate B C, which, being prolonged, would be very nearly tangent at D to the exterior circumference. The lateral boundaries of the canal A B C are prolonged 1 mètre beyond the point C, to prevent the water from falling outside of the wheel. The water passes over the distance C D in virtue of its acquired velocity, and enters the wheel nearly at the top. As quite a narrow opening only is left between the soles of the buckets, the water that flows in the canal A B C is given the form of a thin stratum, by making it pass under a sluice placed near A, and which is raised only about $0^{m}.06$ or $0^{m}.10$. This sluice presents an orifice with rounded edges, so as to avoid the eddies consequent to the exit of the liquid threads. As has been shown, there is little difference in height between the point at which the water enters and the level of the head race; consequently the water enters the wheel with a slight absolute velocity, and if the wheel turn slowly, as it should do, to attain a good effective delivery, the relative velocity will itself be slight, as well as the loss of work that it involves.

The arrangement in (Fig. 9), which has been frequently employed, does not appear to be so good; but we are sometimes obliged to make use of it if the pond level is very variable. This portion is terminated near the wheel by a wooden shutter A B, with openings C C, having vertical faces like those of a window-blind; a movable sluice allows of covering as many of these openings as may be requisite, so as to expend only the disposable volume of water. The inconvenience of this method is, that the water falls through sufficient height into the buckets to give it a considerable increase of velocity; the disturbance of the water in the wheel thus becomes much greater. It tends, moreover, for the same head, to increase the diameter of the wheel, which makes it more heavy and expen-

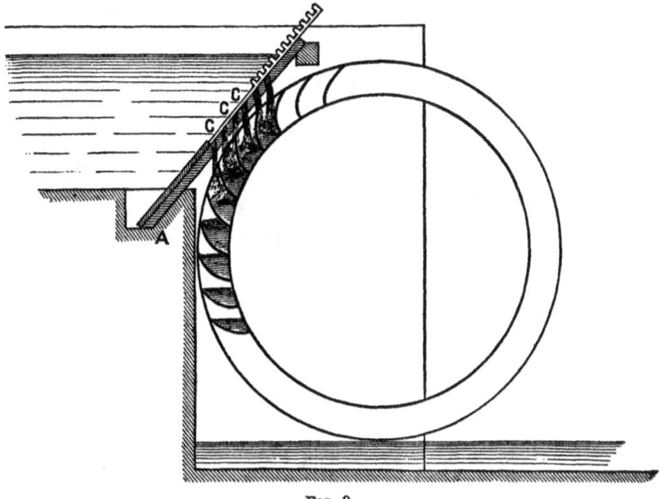

FIG. 9.

sive. Besides, the point at which the buckets take a sufficient inclination to begin to discharge the water in them is situated at a greater height above the lowest point of the wheel, because this height is proportional to the diameter; there is thus, then, a greater loss of head, seeing that the work of the weight of the molecules that have left the buckets, whilst they are falling into the race below, is evidently lost to the wheel.

(*b*) *Shape of the surface of the water in the buckets; velocity of the wheel.*—It can be shown that a heavy homogeneous liquid cannot be in equilibrio relatively to a system that turns uniformly about a horizontal axis. If, however, we admit that the relative equilibrium of the water can exist approximately in the buckets, which may arise when the disturbance due to the entrance of the liquid has nearly ceased, we can determine the shape assumed by the free surface as follows:

Let M (Fig. 10) be a liquid molecule, having a mass m, situated at the distance O M $= r$ from the axis of rotation O; it is in equilibrio relatively with a system which turns around this axis with an angular velocity ω. This equilibrium exists under the action: 1st, of the weight $m g$, which acts vertically along the line M G; 2d, of the centrifugal force $m \omega^2 r$, along the prolongation M C of O M, an apparent force to be introduced, as regards solely a relative equilibrium; 3d, of the pressures produced by the surrounding molecules. We know from the principles of hydrostatics that the resultant of the two first forces is normal to the surface level (or of equal pressure) which passes through M. If, then, M be found at the free surface, as the pressure there is entirely the atmospheric pressure, the resultant in question will be normal to this surface. Let us take M G $= m g$, M C $= m \omega^2 r$, the diagonal M B of the parallelogram M G B C represent the resultant of $m g$ and of $m \omega^2 r$, and consequently it is normal to the free surface. Moreover, prolonging the vertical O A until it intersects this normal, we obtain from the property of similar triangles,

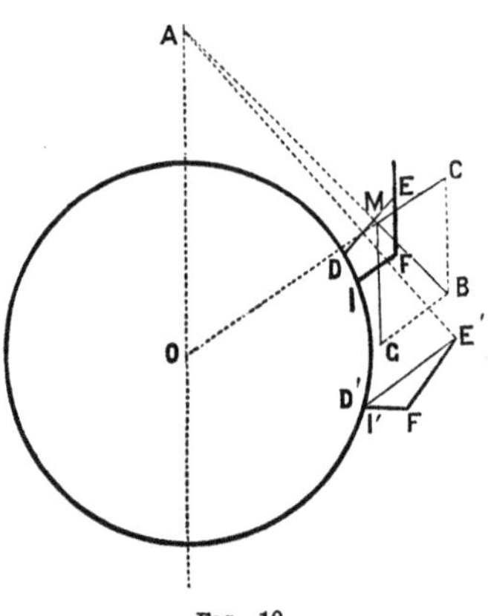

FIG. 10.

$$\frac{\text{O A}}{\text{M G}} = \frac{\text{O M}}{\text{G B}},$$

whence

$$\text{O A} = \frac{\text{M G} \times \text{O M}}{\text{G B}} = \frac{m g r}{m \omega^2 r} = \frac{g}{\omega^2};$$

hence the distance O A is constant, which shows that in a plane section perpendicular to the axis all the normals to the free surface meet at the same point. The profile of the free surface, if there be relative equilibrium, is then necessarily a circle described from the point A as a centre.

This result proves that, in effect, the relative equilibrium is, strictly speaking, impossible; for, in proportion as the bucket leaves its place, the point A not changing position, the free surface would have an increasing radius, which is incompatible with the hypothesis of relative equilibrium, since in this case the form of the free surface ought not to change. The form that has been determined is that which the water endeavors to assume without being able to preserve it.

To finish determining the circle which limits the water in a given bucket, a circle of which we as yet know only the centre A, we must take into consideration the quantity of water that the bucket is to hold. To this end, let N be the number of buckets filled, b the breadth of the wheel parallel to the axis, Q the volume expended by the pond per second. Each bucket occupies on the circumference an angle $\frac{2\pi}{N}$ (expressed in terms of an arc of a circle having a radius 1), and as the wheel turns with an angular velocity ω, $\frac{\omega N}{2\pi}$ will represent the number of buckets filled in a unit of time. Each bucket then contains a volume $\frac{2\pi Q}{\omega N}$, so that the area occupied by the water in the cross section of the bucket has for its value $\frac{2\pi Q}{b\omega N}$. The arc D M E will then be determined, since we know its centre and the surface D M E F I which it must intercept in the given profile of the bucket.

In a certain position I′ E′ F′ of the bucket, the free surface, determined in the way just mentioned, just touches the edge of the exterior side E′; this position may be found by trial. As soon as the bucket passes it, the water begins to run out; for every position below this, it is clear that the free surface will have for a profile a circle having A for a centre and just touching the outer edge of the bucket, and which will allow the volume of water remaining in the bucket to be determined. When the circle in question passes entirely below the profile of the bucket, the discharge will be complete.

In practice, if the question relates to wheels possessing only a slow angular velocity, $\frac{g}{\omega^2}$ will be so great that the circles described from A as a centre, to limit the surface of the water in the buckets, may be assumed as horizontal lines. Example: The wheel being four metres in diameter, and having a velocity of 1 metre at the circumference, then $a = \frac{1}{2}$ and $\frac{g}{\omega^2} = 39^{m}.24$, or the distance of the centre A above the axis.

When an overshot wheel turns rapidly, the distance $\frac{g}{\omega^2}$ may become so small that the free surface may present a noticeable concavity below the horizontal; thus the more the angular velocity increases the less water the bucket can hold in a given position, which is easily seen, since the centrifugal force becomes greater and greater, and this force tends to throw the water out of the bucket. This is an inconvenience attendant upon wheels that turn rapidly; they lose a great deal of water by spilling, and consequently yield a smaller effective delivery.

The losses of head produced by the introduction of the water into the buckets, and by the velocity of the water when it leaves the wheel, also increase with the angular velocity. We will then be led, in order to economize the motive force as much as possible, to make the wheel turn very slowly. But we have already seen, in speaking of breast wheels, that it is not well to make a water wheel move very slow, because, in order to use up an appreciable volume of the water, it would be necessary to establish a machine of immense size. A velocity of from 1 metre to $1^{m}.50$ at the circumference gives good results.

(*c*) *Breadth of the wheel; depth of the buckets in the direction of the radius.*—We have said above that, in a well-arranged wheel, the water leaves the up-stream portion of the troughs by passing under a sluice raised from $0^{m}.06$ to $0^{m}.10$ above the sole; which is itself from $0^{m}.20$ to $0^{m}.25$ below the level of this portion. If we call

Q the expenditure per second;

b the breadth of the wheel, and of the orifice under the sluice;

x the height to which the sluice gate is raised;

h the depth of the sole below the level of the water in the up-stream section;

v the velocity with which the water leaves the sluice;

the velocity v will be due very nearly to the head $h - x$; and as the adjustments are so arranged as to have but little contraction, we can place

$$Q = 0.95\, bx \sqrt{2g(h - x)},$$

the co-efficient 0.95 being intended to account, at a rough estimate, for the loss of head that water always undergoes in any movement whatever, and for the contraction that would yet partially exist. By making, in this expression, $h = 0^{m}.20$, $x = 0^{m}.06$, we deduce $\frac{Q}{b} = 0^{m}.095$; in like manner, for $h = 0^{m}.25$, $x = 0^{m}.10$, we find $\frac{Q}{b} = 0^{m}.163$; that is, with the sluice arranged as we have have said, we can expend from 95 to 163 litres per metre of breadth of the wheel. It would be easy to expend less than 95 litres, by diminishing h and x a little; we can, when necessary, expend more than 163 litres by inverse means. But experience shows that, to be in the best condition, the expenditure should be but little more than 100 litres per metre of breadth; for otherwise we might be led either to make deep buckets, or to cause the wheel to turn rapidly, which would tend to increase the velocity of the water when it enters or leaves the wheel, and consequently to diminish the effective delivery.

To show the relation that exists between the depth p of the buckets in the direction of the radius, and the expense $\frac{Q}{b}$ per yard in breadth, let us preserve the notation already employed in the present number, and furthermore let us call

R the radius of the wheel;

u its velocity at the circumference;

$C = \frac{2\pi R}{N}$ the distance of the buckets apart; c their thickness.

The volume of a bucket will be equal to the product of its three mean dimensions, viz.: its length b, its depth p, and its breadth $C\left(1 - \frac{p}{2R}\right) - c$; this volume has then for its value, $p\,b\,C\left(1 - \frac{p}{2R} - \frac{c}{C}\right)$. Moreover it would be well, to retard the discharge from the bucket, not to have it more than one-third full; the volume of the water it contains would then be $\frac{1}{3} p\,b\,C\left(1 - \frac{p}{2R} - \frac{c}{C}\right)$, and, as we have previously seen, by $\frac{2\pi Q}{\omega N}$, there obtains

$$\frac{1}{3} p\, b\, \mathrm{C} \left(1 - \frac{p}{2\,\mathrm{R}} - \frac{c}{\mathrm{C}}\right) = \frac{2\,\pi\,\mathrm{Q}}{\omega\,\mathrm{N}};$$

whence, because $\mathrm{C} = \frac{2\,\pi\,\mathrm{R}}{\mathrm{N}}$ and $\omega = \frac{u}{\mathrm{R}}$;

$$\frac{\mathrm{Q}}{b} = \frac{1}{3} p\, u \left(1 - \frac{p}{2\,\mathrm{R}} - \frac{c}{\mathrm{C}}\right).$$

As the factor in the parenthesis in the second member differs but little from 1, we may simply place

$$\frac{\mathrm{Q}}{b} = \frac{1}{3} p\, u.$$

This equation shows that when $\frac{\mathrm{Q}}{b}$ is large, one of the factors p or u must be so too. For example, if $\frac{\mathrm{Q}}{b} = 0.100$ litres, and $u = 1$ metre, we find $p = 0^{m}.30$. It is desirable that p should not much exceed $0^{m}.30$. However, if we had an ample supply of water to expend, we might either go beyond this limit, or use a faster wheel, or finally fill the buckets more than one-third.

The expenditure per metre of breadth having been fixed, from what precedes, as much as possible below 100 litres per second, the breadth of the wheel results naturally from the total volume of water to be expended. It is seldom that wheels having a greater breadth than 5 metres are constructed.

(*d*) *Geometrical outline of the buckets.*—The distance of the buckets apart is a little greater than their depth; generally, this last dimension is from $0^{m}.25$ to $0^{m}.28$, and the other about from $0^{m}.32$ to $0^{m}.35$. Their number must be a multiple of the number of arms for facilitating the connections, unless the crown and arms are composed of a single piece.

As to their profile, the annexed outline is frequently made

use of (Fig. 11). After dividing the exterior circumference O A into portions A A′, A′ A″, all equal to the distance of the buckets apart, we take A D $= p$, the depth of the buckets in the direction of the radius, and describe the circumference O D; a third circumference is then drawn, O B, at equal distances from the first two. The radii O A, O A′, O A″ . . .

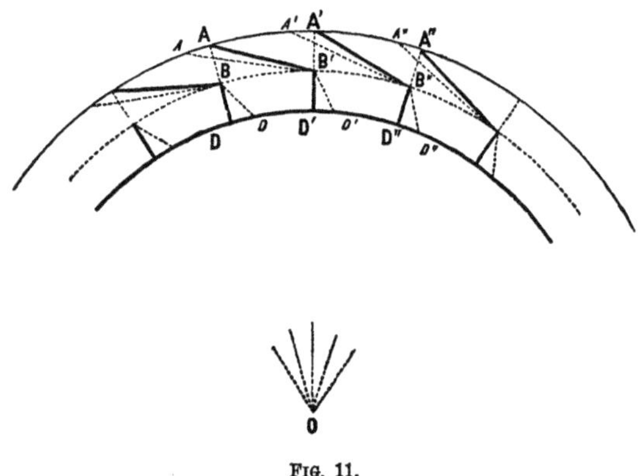

FIG. 11.

being then drawn through the points of division, A B′, A′ B″, . . . will be joined, and we shall then have the profiles A B′ D′, A′ B″ D″, . . . which, excepting the thickness, will be those of the buckets.

Skilful constructors think that, instead of the lines such as A B′, A′ B″, . . . we might employ the lines *a* B′, *a*′ B″, . . which produce a certain degree of mutual covering between the buckets; in like manner for the lines B D, B′ D′, B″ D″, . . . , the inclined right lines B *d*, B′ *d*′, B″ *d*″, . . . , have been sometimes substituted. These two changes have the one end, that of increasing the depth of the buckets in the direction parallel to the circumference, and consequently to retard the emptying. They are inconvenient, because they make the construction more difficult; besides, this overlapping A *a* must not be carried to excess, otherwise the remaining free space between the point B and the side *a* B′ would perhaps be too much diminished. This minimum distance should be a few centimetres greater than the height to which the sluice gate is raised, in order that the water may enter well into the wheel, and not be thrown to the outside.

When the buckets are made of sheet-iron, the broken profiles just mentioned are replaced by curved profiles, which should differ as little as may be from them.

Wooden buckets are generally from 15 to 30 millimetres in thickness; the sheet-iron ones are only from 2 to 4 millimetres, which increases slightly their capacity, all other things being equal. They are limited at the sides by the annular rings, which are fastened to the axle by arms, which increase in number with the diameter of the wheel. They present a continuous bottom or sole throughout the entire circumference D D′ D″. . . . ; in very large wheels this bottom must be sustained by supports at one or two points placed between the exterior crowns. We might also use in this case one or two intermediate crowns.

(*e*) *To calculate the dynamic effect of a head that causes an overshot wheel to turn.*—The two main causes which give rise to the losses of head to be subtracted from the entire head, to obtain the head that is turned to account, are the relative velocity when the water enters the wheel, and that which it possesses at the moment that it falls to the level of the tail race.

It is almost impossible to obtain an accurate value of the first. During the time that a bucket is being filled, the point of entrance of the molecules, which come in successively, is changing in a continuous manner. The first impinge against the solid sides; those that come after, against those that are already in; and thence result phenomena very difficult to analyze. The study is greatly simplified by admitting, as we did in (No. 3), that the height $\frac{w^2}{2g}$, due to the relative velocity w of the water at its point of entrance, represents the loss of head in question. Besides, if we call v the absolute velocity of the water, u the velocity of the wheel, γ the angle formed by the two velocities; as w is the third side of the triangle formed by v and u, we shall have

$$w^2 = u^2 + v^2 - 2\,u\,v \cos \gamma.$$

In reality, the impinging of the water on the wheel takes place at different points along the depth of the bucket. Recollecting now that the radius of the wheel is great compared with the thickness of the shrouding of the buckets, this will not materially affect u; but to determine v and γ, it would be well perhaps to suppose the point of entrance, not at the exterior circumference, but at the middle of the depth of the buckets.

Let us pass to the second loss. Let a molecule of the mass m leave the wheel at a height z above the tail race. This molecule, having only an insensible relative motion in the bucket, possesses, at the moment that it leaves it, the velocity u of the wheel, and at the moment it reaches the level of the tail race it has a velocity v' equal to $\sqrt{u^2 + 2\,g\,z}$. Then it gradually loses all its velocity while moving in this portion, without its piezometric level changing (for we suppose the free surface horizontal in this portion); it undergoes then a loss of head equal to

$$\frac{1}{2\,g}(u^2 + 2\,g\,z) = \frac{u^2}{2\,g} + z.$$

For all the molecules composing the weight P expended in a second, there will be a mean loss expressed by $\frac{1}{\mathrm{P}}\,\Sigma\, m\, g$

$\left(\frac{u^2}{2g} + z\right)$, or else by $\frac{u^2}{2g} + \frac{1}{P} \Sigma\, m\, g\, z$, the sum Σ including all the molecules. This will be the second height to be subtracted from the total height of the head; it is composed of two terms, of which the first is at once given, and it only remains to be seen how we can calculate the term $\frac{1}{P} \Sigma\, m\, g\, z$, which expresses the special effect of the emptying of the buckets.

The quantity $\frac{1}{P} \Sigma\, m\, g\, z$ is nothing more than the mean height comprised between the point of exit of a molecule and the level of the tail race; as the circumstances of all the buckets are exactly the same, it is evidently sufficient to seek this mean for the molecules contained in one bucket. To this end, we will first determine, as stated above (*b*), the positions of the bucket at which the emptying begins and ends, and, for a certain number of intermediate positions, we will ascertain the amount of water that remains in the bucket. Let then

c be the height of the outer edge of the bucket above the level of the tail race when the emptying begins, and let the bucket, still full, hold the volume of water q_0;

c' the analogous height when the emptying has just ended;

y the distance that this same edge has descended whilst the volume of water q_0 was being reduced to q.

During an infinitely small displacement of the wheel, to which the descent $d\,y$ corresponds, an infinitely small volume $-d\,q$ is emptied out, which falls into the tail race from a height $c - y$; the mean height of the outflow will then be $\frac{1}{q_0}\int_0^{q_0} (c-y)\, d\,q$. Now integrating by parts there obtains

$$\int (c-y)\, d\,q = q\,(c-y) + \int q\, d\,y,$$

and, observing that $y = c - c'$ and $y = o$ correspond to the limit $q = o$ and $q = q_0$,

$$\int_0^{q_0} (c - y)\, d\, q = q_0\, c + \int_{c-c'}^{o} q\, d\, y = q_0\, c - \int_o^{c-c'} q\, d\, y;$$

the mean sought is then $\frac{1}{P} \Sigma\, m\, g\, z = c - \frac{1}{q_0} \int_o^{c-c'} q\, d\, y.$

The calculation of the definite integral $\int_o^{c-c'} q\, d\, y$ will be effected by Simpson's method, for want of a strict analysis, since we have the means of determining the value of q corresponding to a given value of y. Were we satisfied with a greater or less approximation, but generally one sufficient, we could, under the sign $\int$, replace the variable q by the mean $\frac{1}{2}\, q_0$ of its extreme values; we should then find,

$$\frac{1}{P} \Sigma\, m\, g\, z = c - \frac{1}{2}(c - c') = \frac{1}{2}(c + c').$$

To the losses already determined we must still add one for bringing the water from the upper portion of the canal to the wheel. As we have seen in examining other wheels, it will be very slight if a good arrangement be adopted; its value would then be $0.1\, \frac{v^2}{2\, g}$.

(*f*) *Practical suggestions.*—The liquid molecules taking, during their fall which follows the discharge, a velocity sensibly vertical, there are scarcely any means for turning this velocity to account by a counter-slope, and consequently the down-stream level should just graze the bottom of the wheel. For wheels that turn rapidly, it would perhaps be advantageous to set them in a mill race, which would only allow the water to escape at the lower portion, and with a velocity nearly horizontal; we should take care to furnish each bucket with a valve, placed near the bottom, and opening from without inwards, to allow the air to enter when the water runs out. It would then be possible to diminish greatly the loss of head occasioned by the velocity that the water possesses on leaving the wheel; but, on the other hand, we would increase, in no small degree, the expense of erecting the wheel, and very likely also that of keeping it in repair.

Overshot wheels answer very well for heads from 4 to 6 metres; less than 3 metres the breast-wheel is to be preferred. Besides, as their diameter is nearly equal to the height of the water fall, their employment would become practically impossible for very high falls.

Experience shows that with an overshot wheel, well set up and moving slowly, the productive force of the head of water may rise as high as 0.80, and sometimes even more. But, in wheels that turn rapidly, it sometimes falls as low as 0.40.

TUB WHEELS.

---

§ III. Water Wheels with Vertical Axles.

10. *Old-fashioned spoon or tub wheels.*—The paddles of spoon-wheels are of slightly concave form in the direction of their length and breadth. They are arranged around a vertical axle, and receive in an almost horizontal direction the shock of a fluid vein which, leaving a reservoir with a great head, is led near the wheel by a wooden trough. To obtain a greater action, the water is made to strike against the concave side of the paddles.

Let us call

v the absolute velocity, supposed to be horizontal and perpendicular to the paddle struck, of the vein which strikes the wheel;

ω the section of this vein;

u the velocity of the paddles at the point at which they receive the shock;

Π weight of the cubic metre of water.

The relative velocity of the water and paddles will be horizontal and equal to $v - u$; then, if this phenomenon be assimilated to that of a fluid vein impinging against a plane, the force exerted on the wheel will be $\Pi\, \omega \frac{(v - u)^2}{g}$; the work which this force performs on the wheel in the unit of time will be expressed by $\Pi\, \omega \frac{u\,(v - u)^2}{g}$, a quantity sensibly equal to the dynamic effect of the head (No. 2). This expression varies with u, and reaches its maximum for $u = \frac{1}{3} v$; this maximum is $\Pi \omega \frac{4}{27} \frac{v^3}{g}$, or, seeing that $\Pi \omega v$ gives the weight P expended per second, $\frac{8}{27} P \frac{v^2}{2g}$. The height $\frac{v^2}{2g}$ can only be a fraction of the head H; hence, the dynamic effect is found below $\frac{8}{27}$ P H, and the effective delivery below $\frac{8}{27}$, or about 0.30. This number, moreover, cannot be considered as strictly correct, because the imperfectness of the theories relating to the resistance of fluids has caused us to give a more or less approximate value for the reciprocal action of the water and the wheel; however, experience confirms the result of calculation, at least inasmuch as it indicates for this class of motors an effective delivery that is always very small, varying from 0.16 to 0.33.

The tub-wheel does not give a much better result. A vertical cylindrical well made of masonry receives the water from the head-race through a conducting channel A (Fig. 12), whose sides incline towards each other with an inclination of about $\frac{1}{5}$ as they recede from the well; that is to say, that the sides make an angle of 11 or 12 degrees with each other.

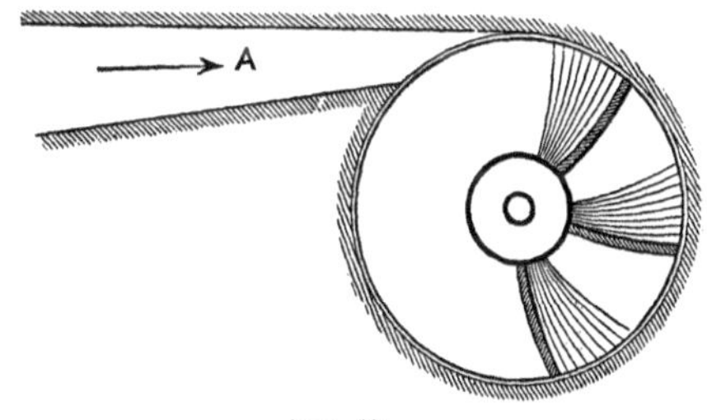

FIG. 12.

One side is also tangent to the circumference of the well. The wheel, whose axis coincides with that of the well, consists of a certain number of paddles regularly distributed around a vertical axle. The horizontal section of the paddles presents a slightly curved form, having its concavity towards the side from which the action of the water comes; cut by a cylinder concentric with the well, they would give inclined lines more or less like arcs of helices. The working of the machine is easily understood: the water comes through the trough A with considerable velocity, endeavors to circulate all around the well, and, meeting the paddles in its road, obliges them to turn, as well as the axle that supports them. At the same time, the water obeys the law of gravity, passes through the wheel by means of the free space between the paddles, and falls into the tail race, which ought to be a little lower. We see that the water must undergo a good deal of disturbance in entering the wheel, and, moreover, that it acts upon the latter for too short a time to entirely lose its relative velocity. Also the effective delivery, sometimes very slight and about 0.15, never exceeds 0.40.

We will pass over these primitive machines in order to study others more perfect

TURBINES.

---

11. *Of turbines.*—The principle of reaction wheels, such as are ordinarily mentioned in Treatises of Physics, has long been known; but it appears that it was only towards the beginning of the last century that the idea was entertained of making use of and applying it to the construction of water wheels with a certain power. Segner, a professor at Göttingen, and more recently Euler, in 1752, made them the object of their researches.

In 1754 Euler constructed another machine, still founded on the principle of reaction wheels, but differing from them in several important arrangements; this machine offers the most striking resemblance to a powerful wheel now in use, called in the arts Fontaine's turbine, from the name of the skilful constructor, who has set up a great many within late years, besides extending and completing, in the details of their application, the idea first advanced by Euler. It appears that this kind of wheel was not much used until towards 1824, the period at which the question was again studied by M. Burdin, engineer in chief of mines, who constructed a similar machine which he called a *reaction turbine.* In the years following, M. Fourneyron, inspired by the ideas of M. Burdin, established some turbines in which he introduced marked improvements. Since that time turbines have greatly spread and multiplied; and there exists a great number of models which differ more or less from each other.

Without undertaking to follow up more completely the history of the changes successively undergone by this kind of wheel, we shall briefly describe the three principal classes into which the turbines now in use may be divided; then we will give a general theory for them, and finally we will mention some details to which a particular interest is attached.

12. *Fourneyron's turbine.*—The essential parts of this turbine are represented in (Fig. 13).

The water from the head race A descends into the tail race B by following a tube, with a circular horizontal section, of which C D is the upper opening. This tube, which is permanently fixed, rests on supports of timber or masonry; it is prolonged by another circular cylinder of cast iron E G I F, movable vertically, which can be lowered more or less, by means to be presently explained. The bottom K K′ K″ L″ L′ L of the tube is joined to a hollow cylinder or pipe *a b c d*, supported at its upper end; this pipe is moreover intended to keep the vertical shaft *e f* from contact with the water, a motion of rotation being given to the shaft by the flow of the water due to its head. In fact, we see that if the water reached the axle, it would be necessary, in order to avoid leakage, to make this latter pass through tightly closed packings, which would occasion friction, independently of that caused by the contact of the fluid. The arbor, as well as the pipe, are, moreover, concentric with the tub.

Between the bottom G I of the cylindrical sluice E G I F and the annular plate K L, there is an opening G K, I L, entirely around the perimeter of the bottom, through which the water can flow. But as it is important, as we shall see, that the threads should not flow in any direction whatever, they are guided in their exit by a certain number of cylindrical partitions with vertical generatrices, which are supported by the

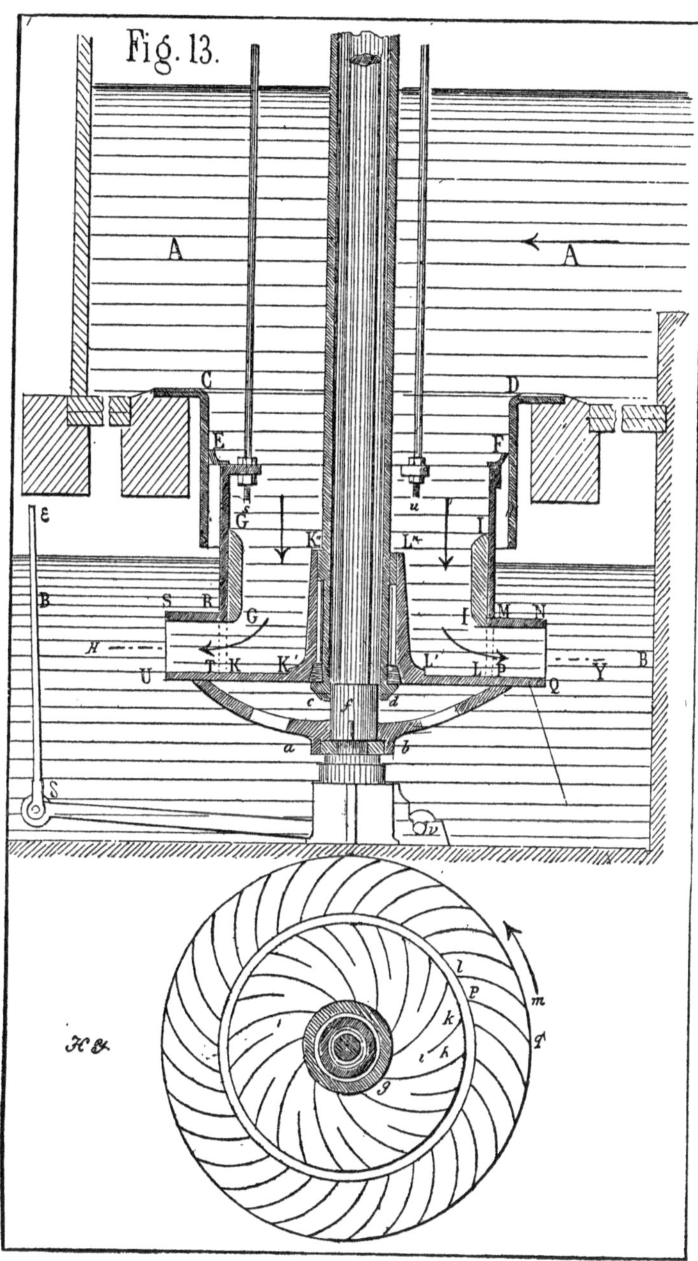

FIG. 13.

plate K L, and of which the arrangement is sufficiently well indicated in the horizontal section; amongst these directing partitions or guides, some, such as $g\,h$, are joined to the sides K′ K″, L′ L″; others, such as $i\,k$, are shorter, in order to avoid too great a proximity in the extremities of the partitions towards the axle.

With regard to the opening G K, I L, the turbine proper is included between the two annular plates or crowns S R M N, V T P Q; these plates are connected together by means of the floats of the turbine, which are cylindrical surfaces with vertical generatrices, giving in horizontal section a series of curves, such as $l\,m$, $p\,q$, &c.; the lower plate is further connected with the arbor by a surface of revolution T $\alpha\,\beta$ P, bolted to it, so as to form a perfectly solid whole. The axle rests on a pivot at its lower end; a lever $\gamma\,\delta$, moved by a rod $\delta\,\varepsilon$, which ends at a point easily reached, allows this pivot to be raised a very little, when the wear and tear of the rubbing surfaces has produced a slight settling of the axle.

To see how the action of the water sets the machine in motion, let us suppose first that the arbor is made fast: then the liquid threads, leaving the well through the directing partitions, will strike against the concavity of the floats; they will thus exert a greater pressure on the concave than on the convex portion of a channel such as $l\,m\,p\,q$, formed of two consecutive floats, first by virtue of the shock, and secondly the curved path that they are obliged to describe. Hence there would result a series of forces whose moments, relative to the axis, would all tend to turn the system of the floats in the direction of the arrow-head indicated in the horizontal section; thus there will be produced effectively a rotation in the direction indicated, if the axle be allowed to turn, even in opposing it by a resistance of which the moment should be inferior to the entire moment of the motive forces.

To diminish as much as possible the loss of head experienced by the liquid molecules during their passage from the head race to the wheel, we should take care: 1st, to give the opening C D a sufficiently great diameter and to round off its edges; 2d, to furnish the circular sluice with wooden appendages G G′, I I′, placed at the lower portion, and having their edges rounded off, as shown in the figure. We shall thus sufficiently avoid the whirls and eddies caused by the successive contractions of the threads of water. The wooden packing, moreover, is not continuous; it is composed of a series of pieces, each occupying the free space between two consecutive partitions, so that the sluice can be lowered without any hindrance as far as the bottom K K′ L L′.

We shall see, in considering the general theory of turbines, how the other conditions essential to a good hydraulic motor are fulfilled.

13. *Fontaine's turbine.*—Fig. 14 shows a general section of this machine by a vertical plane. A pillar or vertical metallic support A B is set as firmly as possible into the masonry forming the bottom of the tail race; this supports at its upper end A a hollow cast-iron axle G D E F, which surrounds it; this axle is prolonged above by a solid one, upon which is the mechanism for transmitting the motion. A screw and nut C allows the position of the axle to be regulated in a vertical direction. Nearly at the level of the water in the tail race (or, if desirable, below it) is placed the turbine H I K L M N O P, permanently fastened to the bottom of the hollow axle; it is comprised between two surfaces of revolution concentric with the vertical axis of the system; these surfaces having H K and I L for meridian lines; in the intermediate space are placed the floats,

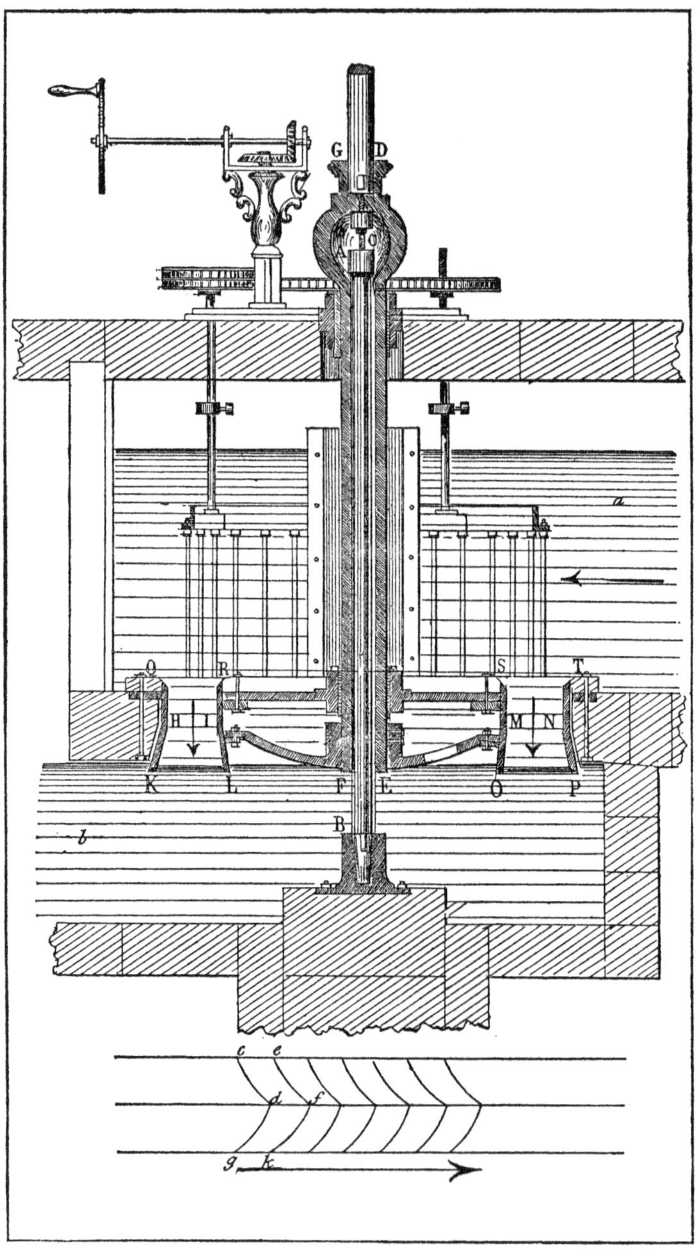

Fig. 14.

which receive the action of the water, and at the same time strengthen the two surfaces. The water comes from the head race *a* to the floats by flowing through a series of distributing canals, of which the quadrilaterals Q R H I, S T M N represent the sections. These channels are distributed continuously over an annular space, directly over the floats; they are limited laterally by the surfaces Q H T N, R I S M; the space between these surfaces, moreover, remains free, except the volume occupied by the directing partitions, which divide it into a certain number of inclined channels, in which the liquid threads move with a determinate figure and direction.

To give a clear idea of the shape of the directing partitions and floats, let us suppose a section made by a cylinder or a cone, concentric with the axis of the system, passing through the middle of the spaces Q R, H I, K L, and this section developed on a plane. The developed section of the directing sections will give a series of curves such as *c d*, *e f*, . . . comprised in a straight or curved row; in like manner, for the floats of the turbine, we shall obtain the curves *d g*, *f h*, . . . also comprised in another row. These curves having been drawn conformably to rules which we shall consider further on, let us suppose reconstructed the cylinder or cone that had been developed, and let us conceive the warped surfaces generated by a right line moving along the axis and on each of the curves in question successively; we shall in this way have determined the surfaces of the partitions and floats.

It is deemed unnecessary to describe the arrangement for the water from the head race to flow into the tail race in no other way than through the channels formed by the directing partitions; in this respect the figure gives sufficiently clear indications.

We can, as in the case of Fourneyron's turbine (No. 12), account for the direction in which the machine should turn on account of the action of the water; and which is that of the arrow-head drawn below the development of the floats.

14. *Kœcklin's turbine.*—Kœcklin's turbine, of which the entire arrangement was first imagined by a mechanic named Jonval, does not differ essentially from Fontaine's turbine, either in the arrangement of the floats and directing partitions, or the mode of action of the water. The most noticeable dif-

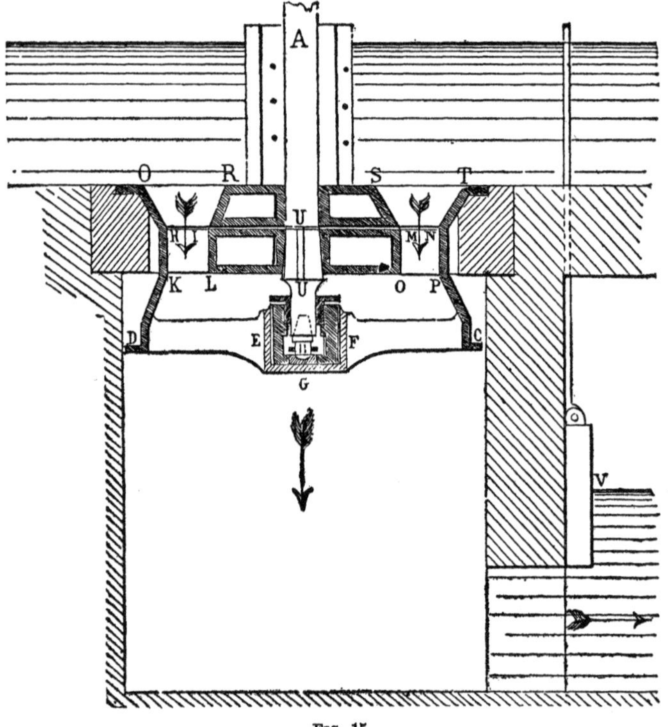

FIG. 15.

ference consists in this, that the turbine is above the level of the water in the tail race, as shown in (Fig. 15), a vertical section of the apparatus. The directing partitions, set in an annular space, of which the trapezoids Q R H I, S T M N, indicate sections, are fastened to a kind of cast-iron socket, which embraces the axle A B, without, however, forming part of it, or pressing it hard; they form a set of inclined channels, through which the water from the head race flows and reaches the floats of the turbine, placed immediately below, as in Fontaine's turbine. These floats are fastened to another socket, which is bolted to the axle; they occupy the annular space H I K L, M N O P.

The inclined channels included between two consecutive floats or partitions are limited on the outside by a fixed tub of cast-iron, resting on the edges of a well of masonry; it forms a surface of revolution about the axis A B, having Q H K D for a meridian section. At the bottom of this tub are found a certain number of arms which support a centre piece, on which is placed the pivot of the revolving shaft.

The water having left the turbine, by the apertures K L, O P, flows into the tail race by descending through the masonry well, and then passing into an opening which we may contract, or, if need be, close at will, by means of a sluice V.

The situation of the turbine above the level of the water in the tail race allows it to be easily emptied, and herein lies its principal advantage; for this purpose we have merely to leave the sluice of the tail race open, and to prevent the water reaching the distributing channels Q R H I, S T M N. We can then visit the machine and make the necessary repairs. Besides, the height included between the horizontal plane K L O P and the level of the water in the tail race should not be considered as a loss of head, because it belongs to a diminution of pressure on the water which leaves the turbine, and we shall see by the general theory, now to be investigated, that this causes an exact compensation.

15. *Theory of the three preceding turbines.*—A complete theory of the turbines that we have just summarily described ought to include first the solution of the following general problem. Having given all the dimensions of a turbine, its position as regards the head and tail races, and finally its angular velocity, to determine the volume of water that it expends, and the dynamic effect of the head, we should then seek the conditions necessary to make the effective delivery for a given head and expenditure of water a maximum.

But we shall not treat the question in such general terms. In order to simplify the researches with which we are to be employed, we shall allow in what follows that all the dimensions have been chosen and the arrangements made, so that the turbine may fulfil in the best possible way the conditions for a good hydraulic motor. Then, from the pond to the exit from the directing partitions, care will have been taken to avoid contractions and sudden changes in direction of the threads; to have smooth and rounded surfaces in contact with the water that flows through, in order that, in this first portion of its passage, it may meet with no sensible loss of head. At the point of entrance into the turbine, the water possesses a certain relative velocity; matters will be so arranged that this velocity shall be directed tangentially to the first elements of the floats, in order that no shock or violent disturbance may follow. This is a condition that it is possible to fulfil by choosing a suitable velocity for the wheel, as well as proper directions for the floats and partitions where they join. Finally, as the water leaves the turbine in every direction about a circumference, and as it is hardly possible to prevent the absolute velocity which it then possesses being used up as a dead loss in producing eddies in the tail race, we shall suppose that we have taken care to make this velocity small. All this combination of circumstances will considerably simplify our calculations, by allowing us to neglect in them, without a very sensible error, the different losses of head experienced by the water up to the point of its exit from the turbine, a loss of which the analytical expression, more or less complicated, would overload our formulas and make them much less manageable. Only, it should be understood that our results will be exclusively applicable to the case in which the machine works according to the conditions for a maximum effective delivery.

Thus granted, let us call

v the absolute velocity of the water when it leaves the directing partitions and enters the turbine;

u its impulsive velocity and w its relative velocity, at the same point, with respect to the turbine taken for a system of comparison;

v', u' and w' the three analogous velocities for the point at which the water leaves the wheel;

p and p' corresponding pressures at these two points;

p_a the atmospheric pressure;

r and r' the distances of the same two points from the axis of rotation of the system*;

H the height of the head, measured between the level of the

(*) These two distances are often equal in the Fontaine and Kœcklin turbines; but they should differ in a marked degree in Fourneyron's turbine. Moreover, it should be observed that, in the Fontaine and Kœcklin turbines, the outlets of the water, from the guide curves, have a certain dimension perpendicular to the axis of rotation; the lengths r and r', be it well understood, should be referred to mean points of these outlets. Thus in Fig. 15, for example, r should be the mean of U M and U N; r', in the same way, should be equal to $\frac{V'O + U'P}{2}$. M N and O P should also be so taken as to be small relatively to r and r', in order that the consideration of a single thread of water only may not cause an appreciable error.

water in the pond and tail race, supposed to be sensibly at rest;

h the depth, positive or negative, of the point of entrance of the water into the interior of the turbine, below the level of the tail race;

h' the height the water descends during its motion in the interior of the turbine, a quantity equal to zero when M. Fourneyron's arrangements are adopted;

Π weight of the cubic metre of water.

Now, all the intervals between two consecutive directing partitions being considered as a first system of curved channels, and all the intervals between two consecutive floats as a second system. For the first system of these channels, represent by,

β the acute angle under which they cut the plane of the orifices that terminate them at the distance r from the axis; this angle β is also that at which the circumference $2 \pi r$ is cut by the partitions;

b the depth or breadth of the orifices of which we have just been speaking, measured perpendicularly to the circumference $2 \pi r$.

For the second system:

θ the angle made by the plane of the orifices of entrance with the direction of the floats, or, what amounts to the same thing, the angle of these floats with the circumference $2 \pi r$, to which we will give, moreover, a direction opposed to the velocity u, the direction of the float being taken in that of the relative motion of the water;

γ the acute angle at which the channels cut their orifices of exit, or, in other words, the angle of the floats with the circumference $2 \pi r'$ about which the above-mentioned orifices are distributed;

b' the depth or breadth of the orifices of exit, measured perpendicularly to the circumference $2 \pi r'$.

The question now is to establish the relations between all these quantities, under the supposition that the conditions of the maximum effective delivery are satisfied. For that purpose we shall first follow the motion of a molecule of water along its path between the two races, and write out the equations furnished by Bernouilli's theorem.

Between a point of departure taken in the pond, where there is no sensible velocity, and the point of exit at the extremity of the directing partitions, there is a head expressed by $H + h + \frac{p_a - p}{\Pi}$; the velocity being v, we have then, under the supposition of a loss of head that can be disregarded between the two points in question,

$$v^2 = 2\,g\,\left(H + h + \frac{p_a - p}{\Pi}\right). \quad . \quad . \quad (1)$$

The water then moves along the floats of the turbine with a velocity at first equal to w and afterwards to w': in this second period, if we call ω the angular velocity of the machine, we know, applying Bernouilli's theorem to the relative motion, there must be added to the real head $h' + \frac{p - p'}{\Pi}$, a fictitious head $\frac{\omega^2\,r'^2 - \omega^2\,r^2}{2\,g}$, or $\frac{u'^2 - u^2}{2\,g}$; still, neglecting the losses of head, which is approximately admissible when the water enters with a relative velocity tangent to the floats, we shall then have to place

$$w'^2 - w^2 = 2\,g\,\left(h' + \frac{p - p'}{\Pi}\right) + u'^2 - u^2 \quad . \quad . \quad . \quad . \quad (2)$$

When the turbine is immersed and h is positive, the point of exit of the water is found at a depth $h + h'$ below the level of the tail race; besides, as, for the maximum effective delivery, the water must go out with a slight absolute velocity, we can without material error admit that the pressure varies, in the tail race, according to the law of hydrostatics, which gives

$$\frac{p_a}{\Pi} + h + h' = \frac{p'}{\Pi} \quad . \; . \; . \; . \; (3)$$

This relation is also true in Kœcklin's turbine, although $h + h'$ becomes negative, provided that the well placed below the turbine and the orifice by which it communicates with the tail race are sufficiently large; for then the water will take up in it but a slight velocity and can there be considered as in equilibrio; the pressure p' would then be less than p_a by a quantity represented by the depth $-(h + h')$, as equation (3) shows. We could also preserve it, if the lower plane of the turbine, constructed according to one of the first two systems, were on the same level as the water in the tail race: we would then, in fact, have $p_a = p'$ and $h + h' = o$, the quantities h and h' being with contrary signs, or else both equal to zero, according to whether we are considering Fontaine's or Fourneyron's turbine.

The incompressibility of water will furnish us with the fourth equation, showing that the volume of water that has flowed between the directing partitions is equal to that which leaves the turbine. The distributing orifices, left free between the partitions, occupy a total development $2 \pi r$ (excepting the slight thickness of the partitions) and a breadth b, from whence there results a surface $2 \pi b r$; as they are cut by liquid threads moving with the velocity v, at an angle β, we have for the first expression of the volume Q expended in a unit of time

$$Q = 2 \pi b r \sin \beta . v.$$

In like manner, the orifices of exit at the extremity of the floats have a total development $2 \pi r'$, a breadth b', a surface $2 \pi b' r'$, and they are cut by the threads moving with a relative velocity w' at an angle γ'; then

$$Q = 2 \pi b' r' \sin \gamma . w'.$$

Strictly speaking, on account of the thickness of the floats or partitions, these two expressions for the value of Q ought to undergo a slight relative reduction of $\frac{1}{25}$ or $\frac{1}{30}$; but in all cases, the reduction being the same for both, we shall have by the equality of the values of $\frac{Q}{2\pi}$

$$b\,r\,v \sin \beta = b'\,r'\,w' \sin \gamma. \quad . \quad . \quad . \quad . \quad (4)$$

The three following relations are in a certain degree geometrical.

Let us represent (Fig. 16) a float B C and a directing partition A B; a liquid molecule having followed the path A B arrives at B with an absolute velocity v, and a velocity w relatively to the turbine which itself, at the point B, possesses the velocity u. This last being what is called the propelling velocity, we know that v is the diagonal of the parallelogram constructed on u and w; and as the angle between v and u is exactly β, the triangle B U V will give

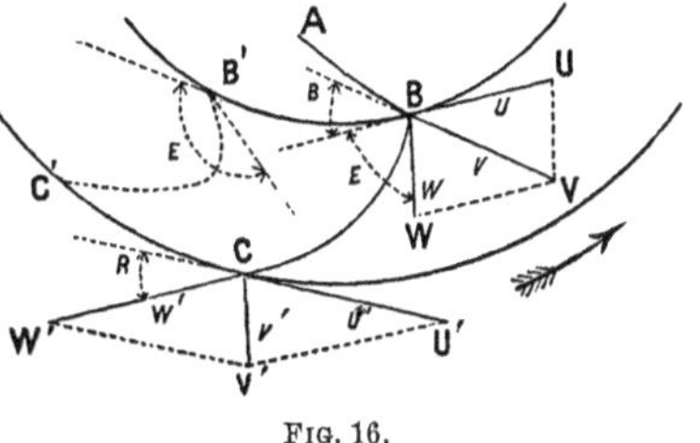

Fig. 16.

$$\overline{UV}^2 = \overline{BU}^2 + \overline{BV}^2 - 2\,\overline{BU}.\overline{BV} \cos \beta,$$

that is,

$$w^2 = u^2 + v^2 - 2\,u\,v \cos \beta. \quad . \quad . \quad . \quad . \quad (5)$$

In like manner the liquid molecule, after having traversed, relatively to the turbine, the path B C, arrives at C with the velocity w', which, taken as a component with the propelling velocity u', gives the absolute velocity v'; then the angle γ being the supplement of that made by u' and w', we shall have

$$v'^2 = u'^2 + w'^2 - 2\,u'\,w' \cos \gamma. \quad . \quad . \quad . \quad . \quad (6)$$

On the other hand, the velocities u and u' belong to two points of the turbine situated respectively at the distances r and r' of the axis of rotation, we have then $\frac{u}{r} = \frac{u'}{r'}$, or

$$u' r = u r'. \quad . \quad . \quad . \quad (7)$$

There still remains to express two conditions necessary for obtaining the best effective delivery. It is necessary first, at the point B, that w be directed tangentially to the floats B C without which there would be a sudden change of relative velocity, whence would result disturbance and a loss of head that we have not considered. Now the angle between w and u is the supplement of θ, hence the triangle B U V gives

$$\frac{\text{B U}}{\text{B V}} = \frac{\sin \text{B V U}}{\sin \text{B U V}} = \frac{\sin (\text{B U V} + \text{V B U})}{\sin \text{B U V}},$$

or

$$\frac{u}{v} = \frac{\sin (\theta + \beta)}{\sin \theta}. \quad . \quad . \quad . \quad (8)$$

It is then necessary that the absolute velocity v' possessed by the water on leaving the turbine should be very slight, since $\frac{v'^2}{2g}$ enters in the loss of head (No. 3): this condition is sufficiently satisfied by taking the angle γ small, and placing

$$u' = w'; \quad . \quad . \quad . \quad (9)$$

for then the parallelogram C U′ V′ W′ is changed into a lozenge, very obtuse at one angle and very acute at the other, and the diagonal joining the obtuse vertices is short; in other words, the velocities u' and w' are equal, and almost directly opposed, which makes their resultant very small.

We have thus obtained, in all, nine equations between sixteen variable quantities in a turbine, namely—

six velocities u, v, w, u', v', w',

two pressures p, p',

two ratios $\frac{r}{r'}$, $\frac{b}{b'}$,

three altitudes H, h, h',

three angles β, γ, θ.

These equations will serve us for solving two distinct problems, which may be thus stated: 1st, having given a turbine and all its dimensions (that is to say, the eight quantities β, γ, θ, $\frac{r}{r'}$, $\frac{b}{b'}$, H, h, h'), to show the conditions these dimensions must satisfy, in order that the turbine may work with the maximum effective delivery—that is to say, so that the nine above equations may obtain; and, under the supposition that these conditions are fulfilled, to show the most suitable velocity of the turbine, as well as the expenditure of water corresponding to this velocity, its effective delivery, and its dynamic effect; 2d, having given the expenditure and the height of a head, to establish under this head a turbine with the best conditions.

The first question involves eight unknown quantities, which are u, v, w, u', v', w', p, p'; the elimination of these unknown quantities between the nine equations will then give an equation of condition to be satisfied by the dimensions of the machine—an equation to which we shall have to add two others in order to show that the pressures p and p' are essentially positive. The following calculation has for its object to bring out these three conditions, and, at the same time, to give the value of the unknown quantities.

Adding equations (1), (2), and (5), member to member, there obtains

$$w'^2 = 2\,g\left(\mathrm{H} + h + h' + \frac{p_a - p'}{\Pi}\right) + u'^2 - 2\,u\,v\cos\beta,$$

or, considering (3) and (9),

$$u\,v\cos\beta = g\,\mathrm{H} \quad . \; . \; . \; . \quad (10)$$

The combination of equations (4) and (9) readily give

$$b\, v\, r \sin \beta = b'\, u'\, r' \sin \gamma\,; \quad . \ . \ . \ . \ (11)$$

multiplying equations (7), (10), and (11), member by member, there obtains

$$v^2\, b\, r^2 \sin \beta \cos \beta = g\, \mathrm{H}.\, b'\, r'^2 \sin \gamma\,;$$

whence one of the unknown quantities

$$v^2 = g\, \mathrm{H} \frac{b'\, r'^2}{b\, r^2} \frac{\sin \gamma}{\sin \beta \cos \beta} \quad . \ . \ . \ . \ (12)$$

We have besides, from equation (10), $u^2 = \dfrac{g^2\, \mathrm{H}^2}{v^2 \cos \beta}$; whence

$$u^2 = g\, \mathrm{H} \frac{b\, r^2}{b'\, r'^2} \frac{\tan \beta}{\sin \gamma}, \quad . \ . \ . \ . \ (13)$$

and from equation (7)

$$u'^2 = g\, \mathrm{H}\, \frac{b}{b'} \frac{\tan \beta}{\sin \gamma}. \quad . \ . \ . \ . \ (14)$$

To obtain v', we will first make $w' = u'$ in equation (6), which will give

$$v'^2 = 2\, u'^2\, (1 - \cos \gamma),$$

and consequently from the value of u' (14),

$$v'^2 = 2\, g\, \mathrm{H}\, \frac{b}{b'} \frac{\tan \beta}{\sin \gamma} (1 - \cos \gamma) \quad . \ . \ . \ . \ (15).$$

Knowing u' we obtain w', and, if w were required, we could easily obtain it by substituting in equation (5) the values of v and u. Thus, all the velocities may be considered as known; we could deduce from them the angular velocity ω with which the turbine must move, when it is working with the conditions of the maximum effective delivery; we would then have practically

$$\omega = \frac{u}{r} = \frac{u'}{r'}.$$

The corresponding expense Q has for its value $2\, \pi\, b'\, w'\, r' \sin \gamma$, or, substituting in place of w' the value of its equal u' from equation (14),

$$Q = 2\pi r' \sqrt{b b'} \sqrt{g H \tan\beta \sin\gamma}, \quad . \; . \; . \; . \quad (16)$$

a formula whose second member we should probably have to multiply by a number less than unity, in order to take into consideration the space occupied by the floats, and also to compensate for the influence of the losses of head neglected in the calculation.

Let us now seek the three equations of condition to be satisfied by the dimensions of the turbine. First, by dividing equations (13) and (12), member by member, and extracting the square root of the quotient, we will find

$$\frac{u}{v} = \frac{b\, r^2}{b'\, r'^2} \frac{\sin\beta}{\sin\gamma},$$

and by reason of eq. (8)

$$\frac{\sin(\theta + \beta)}{\sin\theta} = \frac{b\, r^2}{b'\, r'^2} \frac{\sin\beta}{\sin\gamma}; \quad . \; . \; . \; . \quad (17)$$

this is the condition obtained by eliminating eight unknown quantities between nine equations. There remains yet to express that $p > 0$, $p' > 0$. As to this last condition, we see from (3) that it is itself satisfied for Fourneyron's and Fontaine's turbines, by supposing that they are on a level with the water in the tail race or below it, as we have admitted in the preceding calculations; because then $h + h'$ is positive, and we have $p' > p_a$. In Kœcklin's turbine, on the contrary, the bottom of the turbine is in reality above the level of the water in the tail race, by a positive height expressed by $-(h + h')$; $\frac{p'}{\Pi}$ has for its value $\frac{p_a}{\Pi}$, or $10^m.33$ less this height; then it is absolutely necessary to have

$$-(h + h') < 10^m.33,$$

and perhaps even, on account of neglected losses of head, it would be well to place

$$-(h + h') < 6^{m}., \quad . \quad . \quad . \quad . \quad (18)$$

in order to make perfectly sure of the continuity of the liquid column in the cylindrical well, above which the turbine is found. As to the pressure p, it will be found from eq. (1) after substituting in it for v^2 its value in eq. (19), that is

$$\frac{p}{\Pi} = \mathrm{H} + h + \frac{p_a}{\Pi} - \frac{1}{2}\,\mathrm{H}\,\frac{b'\,r'^2}{b\,r^2}\,\frac{\sin\gamma}{\sin\beta\cos\beta}. \quad . \quad . \quad . \quad . \quad (19)$$

The second member of this equation should, of course, be greater than zero; but we may assign it a higher limit. In fact, if we examine the arrangement of the different systems of turbines, we see that there is always an indirect communication between the distributing orifices, situated at the end of the directing partitions, either with the tail race, or with the external air. This communication is effected by the play necessarily left between the turbine proper and the distributing orifices. When it takes place with the tail race, p cannot differ much from the hydrostatic pressure $p_a + \Pi\,h$, which would take place in a piezometric column communicating with this race, and at the height of the point of entrance of the water above the turbine; otherwise there would be, on account of the play that we have just spoken of, either a sudden gushing out, or suction of the water, which would produce a disturbance in the motion. When it is with the atmosphere, p must, for a similar reason, be equal to p_a. It is then prudent, in the first case, to impose the condition that the two terms in which the factor H appears, eq. (19), should nearly cancel each other, or, designating by k a number that differs little from unity, to make

$$k = \frac{b'\,r'^2}{b\,r^2}\,\frac{\sin\gamma}{\sin\beta\cos\beta}; \quad . \quad . \quad . \quad . \quad (20)$$

k is moreover rigorously subjected to the condition that

$$h + \frac{p_a}{\Pi} + \mathrm{H}\,(1 - k)$$

should be positive. And in like manner, for the second case, we should establish the condition

$$H\left(1 - \frac{b' r'^2}{b r^2} \frac{\sin\gamma}{2\sin\beta\cos\beta}\right) + h = h'' \quad \ldots . \quad (20 \text{ bis})$$

h'' being a very small height.

Again it might be proposed, for a turbine known to be working with the maximum effective delivery, to seek this delivery as well as the dynamic effect. As we are supposing that all losses of head, other than that due to the velocity of exit v', may be disregarded, the head that is turned to account will be

$$H - \frac{v'^2}{2g},$$

and consequently the productive force μ will be expressed by

$$\mu = \frac{H - \frac{v'^2}{2g}}{H} = 1 - \frac{v'^2}{2gH},$$

or replacing v'^2 by its value

$$\mu = 1 - \frac{b}{b'} \frac{\tan\beta}{\sin\gamma}(1 - \cos\gamma). \quad \ldots . \quad (21)$$

The dynamic effect T_e would be obtained by finding the product $\mu\, \Pi\, Q\, H$ of the effective delivery by the absolute power of the head; then we would have, from eqs. (16) and (21)

$$\left.\begin{array}{l} T_e = \Pi H \sqrt{gH}.\ \ 2\pi r' \sqrt{b b'}.\ \ \sqrt{\tan\beta\sin\gamma} \\ \times \left\{1 - \frac{b}{b'} \frac{\tan\beta}{\sin\gamma}(1 - \cos\gamma)\right\} \end{array}\right\} \ldots . \quad (22)$$

Thus have we now solved the first of the two general problems proposed. When we take up the second, which consists in setting up a turbine for a given head, Q and H become the known quantities, and we have, between the nine quantities $\beta, \gamma, \theta, r, r', b, b', h, h'$, which define the unknown dimensions, only equations (16), (17), (20), or (20 *bis*), to which must be added (if a Kœcklin's turbine is in question) the inequality (18); still this inequality leaves a certain margin; and it is the same with equations (20) and (20 *bis*), because the quantities k and h'' have not a definite value. It appears, then, that the problem is very indeterminate, and that we may assume almost all the above-mentioned dimensions arbitrarily; however, the following remarks impose restrictions that it will be well to keep in mind.

16. *Remarks on the angles* β, γ, θ, *and on the dimensions* b, b', r, r', h, h'.—If we only considered the expression for the effective delivery eq. (21), we should be tempted to make one of the angles β or γ equal to zero; the theoretical effective delivery would then become practically equal to unity. But we see that the expenditure Q would reduce to zero, as well as the dynamic effect T_e: hence the value zero is not admissible for either of these angles.

Making γ very small, the channels formed by two consecutive floats would be very much narrowed at the point of exit of the water; the water would flow with difficulty through these narrow passages, and there would be danger of its not following exactly the sides of the floats, which would occasion eddies and losses of head. On the other hand, a large value for γ would diminish, very likely, the effective delivery too much. Between these two points to be avoided, experience gives a value of 20 or 30 degrees as affording satisfactory results.

As to the angle β, besides the reason already given, there is still another for not making it zero: this second reason is that from equation (19) p would be negative for $\beta = o$ and $\beta = 90°$. We must not then approach too closely to zero or to 90°; the limits from 30 to 50 degrees have been advised by some experts; but there is none that is absolute.

Let us suppose that we are about to apply equation (20); multiplying it, member by member, by equation (17) we find

$$\frac{k \sin(\theta + \beta)}{\sin \theta} = \frac{1}{2 \cos \beta},$$

whence

$$\frac{1}{k} - 1 = \frac{2 \cos \beta \sin(\theta + \beta) - \sin \theta}{\sin \theta}$$

or, by developing the sin $(\theta + \beta)$,

$$\begin{aligned} 2 \cos \beta \sin(\theta + \beta) - \sin \theta &= \sin \theta (2 \cos^2 \beta - 1) + 2 \sin \beta \cos \beta \cos \theta \\ &= \sin \theta \cos 2\beta + \cos \theta \sin 2\beta \\ &= \sin(2\beta + \theta); \end{aligned}$$

we can then write

$$\frac{\sin(2\beta + \theta)}{\sin \theta} = \frac{1}{k} - 1. \quad . \quad . \quad . \quad . \quad (23)$$

We have previously seen that k should be a number very near unity; it follows that sin $(2\beta + \theta)$ should be small, and consequently that $2\beta + \theta$ should differ but little from 180 degrees. If, for example, we took β about 45 degrees, θ would be about a right angle. Besides, it is not well to have θ greater than 90°; for, if we refer to Fig. 16, we see that if θ be obtuse the floats should have a form like B′ C′, presenting a considerable curvature; and experience shows that in a very much curved channel the water meets with a greater loss of head, all other things being equal: the liquid molecules then tend to separate from the convex portion, which gives rise to an eddy. We see also in Fig. 16 that, in taking θ very acute, the side V U of the triangle B U V—that is, the relative velocity at the entrance, would tend to become more or less great, which would be hurtful, since the friction of the water on the floats would be increased. Hence θ should be an acute angle, but at the same time almost a right angle: we might make it vary, for example, between 80 and 90 degrees.

If it be equation (20 *bis*) and not equation (20) that we have to apply, the same reasons obtain for taking θ acute and nearly 90 degrees; but the sum $2\beta + \theta$ need no longer differ much from 180 degrees.

After having fixed the values of β, γ, and θ, we will find from equation (17) the ratio $\frac{b\, r^2}{b'\, r'^2}$, whence we can obtain either $\frac{b}{b'}$ or $\frac{r}{r'}$, the other having been assumed.

It is in favor of the effective delivery to have $\frac{b}{b'}$ less than unity, as formula (21) shows; it must not, however, be greater than the difference $b' - b$, and must be proportional to the length of the floats, in order that the channels between two consecutive floats may not be emptied too rapidly, because this emptying would give rise to a loss of head. We can impose the condition that $b' - b$ should be less than $\frac{1}{10}$ the length of the floats.

As we have already said, the ratio $\frac{r'}{r}$ is often taken equal to unity in Fontaine's and Kœcklin's turbines, but it is necessarily greater than unity in Fourneyron's turbine. If it be taken different from unity, we must not, except for particular reasons, increase the difference $r' - r$, or $r - r'$, for we should thus lengthen the floats and increase friction. In Fourneyron's turbine $\frac{r'}{r}$ varies ordinarily between 1.25 and 1.50.

The height h', from which the water descends into the interior of the turbine, is always zero in Fourneyron's turbines; in the other two systems it is so taken that the floats may be sufficiently, but not too long, regard being had to the difference $b' - b$. As to the height h, if it be a question of a turbine of

Kœcklin's, it is fixed according to local circumstances, the inequality expressed in (18) being considered; if it be of one of Fourneyron's or Fontaine's, we so arrange matters that its lower plane may be on the same level as the waters in the tail race when at their minimum depth.

Finally, M. Fourneyron recommends giving the circular section of the tub, in which the directing partitions of his turbines are placed, a surface at least equal to four times the right section of the distributing orifices, in order that the fluid threads may easily pass from the vertical to the horizontal direction, which they must have at their point of exit. With the notation employed in (No. 15), we can write

$$\pi r^2 > 4 . 2 \pi r b \sin \beta,$$

or else

$$r > 8 b \sin \beta, \quad . \quad . \quad . \quad . \quad (24)$$

the sign $>$ not being exclusive of equality.

Let us now show, by two examples, how we shall be enabled, by means of these considerations, to determine the dimensions of a turbine to be established.

17. *Examples of the calculations to be made for constructing a turbine.*—Let it first be determined to establish a Fourneyron turbine with the following data:

Height of fall, $H = 6^{m}.00$.

Volume expended per second, $Q = 1^{mc}.50$.

The absolute power of the head is $1500 \times 6^{kgm} = 9000$ kilogrammètres per second, or 120 horse-power.

Since the angle γ is not fixed theoretically, we will take it (No. 16) equal to 25 degrees; we will also make $k = 1$ (*), which secures that p shall be positive (No. 15); finally, we will take $\theta = 90°$. Equation (23) then gives

$$\sin (2 \beta + \theta) = 0,$$

whence

$$2\beta + \theta = 180^\circ \text{ and } \beta = 45^\circ.$$

As we have satisfied equation (23), which results from the elimination of $\frac{b' r'^2}{b r^2} \frac{\sin\gamma}{\sin\beta}$ between formulæ (17) and (20), it is sufficient to preserve one of these last; we deduce from both

$$\frac{b\, r^2}{b'\, r'^2} = \sin 25^\circ = 0.4226. \quad . \quad . \quad . \quad . \quad . \quad . \quad . \quad . \quad . \quad . \quad . \quad . \quad (\alpha).$$

The condition of expending $1^{\text{mc}}.50$ is expressed by formula (16), which here becomes

$$1^{\text{mc}}.50 = 2\pi r' \sqrt{b b'} \sqrt{6 g.\ 0.4226},$$

or else, by reducing

$$r' \sqrt{b b'} = 0.04785. \quad . \quad . \quad . \quad . \quad . \quad . \quad (\alpha').$$

We have still to express the inequality (24), which gives

$$r > 8\, b \sin 45^\circ \text{ or } r > 5.657\, b;$$

we will take

$$r = 6\, b. \quad . \quad . \quad . \quad . \quad . \quad . \quad . \quad . \quad (\alpha'')$$

We have thus only three equations involving b, b', r, r'; but on account of their particular form we may still deduce the values of b and r. Extracting the square root of equation (α) and multiplying it member by member by (α'), we make $r' \sqrt{b'}$ disappear and find

$$b\, r = 0.031107,$$

a relation which, combined with $r = 6\, b$, gives

$$r = 0^{\text{m}}.432, \qquad b = 0^{\text{m}}.072.$$

This being done, the system of the three equations (α), (α'), (α'') would no longer give anything but $r' \sqrt{b'}$; to avoid any indetermination, we will take b' arbitrarily and deduce r', except that the conditions mentioned in (No. 16), and not thus far expressed, must subsequently be verified. If we take, for example, $b' = 0^{\text{m}}.090$, equation (α') will become

$$r' \sqrt{0.090 \times 0.072} = 0.04785,$$

whence we obtain

$$r' = 0^{m}.594.$$

These values of b' and r' may be retained, because $\frac{r'}{r} = 1.37$, and the difference $b' - b = 0^{m}.018$ is only $\frac{1}{9}$ of $r' - r$, a quantity which, on account of the obliquity of the floats to the exterior circumference, should be but little greater than two-thirds of the length of these last; the discharge of water will not then be too rapid.

The height h now alone remains to be determined: if the level of the tail race were constant, we should make $h = o$; however, we would have to consider what is said (No. 16) on this subject.

The theoretical effective delivery will be obtained from formula (21); we find

$$\mu = 1 - \frac{0.072}{0.090}. \operatorname{tang} 45^{\circ}. \frac{1 - \cos 25^{\circ}}{\sin 25^{\circ}} = 0.823.$$

In practice, we only rely upon a net effective delivery of from 0.70 to 0.75 at most; this it is well to do, on account of all the losses of head that we have neglected, and also because it is very difficult to make a machine move exactly with the velocity and expenditure of water corresponding to the maximum effective delivery.

Finally, to obtain the velocity with which the turbine should revolve, formula (14) should be applied, and we would deduce therefrom $u' = 10^{m}.555$, since the angular velocity $\omega = \frac{u'}{r'} = 17.77$, and finally the number of revolutions per minute $N = \frac{30\,\omega}{\pi} = 169.7$.

Again, let it be proposed as a second example to set up one

of Fourneyron's turbines, with a head of 2 metres, expending $0^{mc}.60$ of water per second, which corresponds to an absolute work of 1200 kilogrammètres, or about a 16 horse-power. We will suppose that the water in the tail race only rises to the level of the lower plane of the turbine, so that the interval or play between the turbine and the directing partitions may communicate directly with the atmosphere, and that the pressure p is sensibly equal to the atmospheric pressure. We will then assume equation (20 *bis*), making in it $h'' = o$, in other words we will place

$$\mathrm{H}\left(1 - \frac{b' r'^2}{b r^2} \frac{\sin \gamma}{2 \sin \beta \cos \beta}\right) + h = o.$$

Following the ordinary rules, we will make $r = r'$; besides, it may be remarked that, on account of the position assigned to the plane of the tail race, h is equal to h' with a contrary sign. The above equation can then be written

$$\mathrm{H}\left(1 - \frac{b'}{b} \frac{\sin \gamma}{2 \sin \beta \cos \beta}\right) - h' = o. \quad . \quad . \quad . \quad . \quad . \quad (\delta)$$

As θ can only differ slightly from 90 degrees, we will give it this value; the equation of condition (17) then takes the form

$$\frac{b\, r^2}{b' r'^2} \frac{\tan \beta}{\sin \gamma} = 1;$$

and, because $r = r'$

$$b' \sin \gamma = b \tan \beta. \quad . \quad . \quad . \quad . \quad . \quad . \quad . \quad (\delta)$$

Introducing $b' \sin \gamma$ in the place of $b \tan \beta$ in the expression (16) of the expenditure, it becomes

$$\mathrm{Q} = 2 \pi r' b' \sin \gamma \sqrt{g \mathrm{H}}. \quad . \quad . \quad . \quad . \quad . \quad (\delta'')$$

The equations (δ, δ', δ'') are those pertaining to the problem. They contain six unknown quantities, viz.: $\beta, \gamma, b, b', r', h'$; we consequently see that they are indeterminate, and that we can assume three of the unknown quantities or three new equations. The angle γ not being determinable by theory, we will take it at first equal to 30 degrees (No. 16); equation (δ'') will become, by substituting numbers for letters,

$$b' r' = \frac{0.60}{\pi \sqrt{2g}} = 0.04312,$$

a relation which is satisfied by the values

$$r' = 0^m.60, \qquad b' = 0^m.072.$$

We thus see that the ratio $\frac{b'}{r'}$ is only 0.12; consequently the inequality in the velocities of the liquid threads in the orifice of exit will not be too noticeable. Now as there remain three unknown quantities, b', h', β, connected only by the two equations (δ), (δ'), we will assume $h' = 0^m.15$; then eliminating $\frac{b'}{b}$ between (δ) and (δ'), we will obtain

$$1 - \frac{1}{2\cos^2\beta} = \frac{h'}{H} = \frac{3}{40},$$

whence

$$2\cos^2\beta = \frac{40}{37}, \ 2\cos^2\beta - 1 = \cos 2\beta \ \frac{3}{37},$$

and consequently

$$\beta = 42^\circ\ 40 \text{ very nearly.}$$

Knowing β, we obtain from equation (δ')

$$b = 0^m.039.$$

The difference $b' - b = 0^m.033$ is perhaps too great relatively to the height $0^m.15$ of the turbine, because the floats having a development of from $0^m.20$ to $0^m.25$ at the most, their spread would reach the amount of $\frac{1}{7}$ very nearly. We then try another value of h'; for example, let

$$h' = 0^m.30.$$

Proceeding as before, we will have successively

$$2\cos^2\beta = \frac{20}{17}, \ \cos 2\beta = \frac{30}{17}, \ \beta = 40^\circ, \ b = 0^m.043.$$

The difference $b' - b$ would then be $0^m.029$; but as the floats would be nearly $0^m.40$ in length (on account of their inclination to the lower plane of the turbine), this number appears perfectly admissible. We should then obtain the results

$$\gamma = 30°, \beta = 40°, \theta = 90°, r = r' = 0^m.60$$
$$b = 0^m.043, b' = 0^m.072, -h = h' = 0^m.30.$$

The effective delivery μ would be given by equation (21), which, considering equation (δ'), becomes

$$\mu = \cos \gamma = \cos 30° = 0.866.$$

In like manner equation (14) would be simplified and give

$$u'^2 = g\,\text{H},$$

whence

$$u' = 4^m.429,$$

from this we deduce finally the angular velocity to be given to the machine

$$\omega = \frac{u'}{r'} = 7.382,$$

and the number of revolutions per minute

$$\text{N} = \frac{30\,\omega}{\pi} = 70.5.$$

In general, as we have seen, the problem which consists in fixing the dimensions of a turbine for which we have the expenditure and the head is indeterminate; we take advantage of this to assume in part the unknown dimensions, and by the method of trial, if that be necessary, endeavor to satisfy the different conditions to be fulfilled, but which the equations do not express.

18. *Of the means of regulating the expenditure of water in turbines.*—A turbine constructed with assigned dimensions, in order to move with the maximum effective delivery, should expend a perfectly determinate volume of water, so long at least as the head remains constant. However, in practice, we are

obliged to regulate the expenditure by the volume furnished by the pond; for if we expended more we would be exposed to the lack of water, after some time, and we would then be forced to suspend the operation of the machine. Consequently we so calculate the dimensions as to expend suitably the greatest volume of water at our disposal, and we make proper dispositions to expend less when the supply diminishes. For this purpose several methods have been employed.

In Fourneyron's turbine, the movable tub E G F I (Fig. 13) allows this end to be attained: it is merely necessary to lower it more or less, to contract the openings G K, I L, or to close them entirely. The vertical motion of translation of this tub is obtained by means of three vertical rods, as *r s*, *t u*, which are attached to it at three points, which form the vertices of an equilateral horizontal triangle. These rods are terminated at their upper ends by screws, and enter into nuts which are forced by their construction to turn in their places. The three nuts, moreover, are furnished with three cogged wheels, all exactly alike, which are geared on the same wheel, which is loose on the axle of the turbine. By turning one of the nuts by means of a wrench, the other two are turned exactly the same amount, and the tub is raised or lowered by the three rods at once.

There is one great inconvenience attendant upon the partial obstruction of the openings G K, I L; it is that the fluid veins which issue through these openings immediately enter canals of greater section, in which they flow necessarily through a full pipe, since the turbine is below the tail race. There is a sudden change of section thus produced, and consequently a greater or less loss of head. This influence is sometimes so great that General Morin has mentioned, in different experiments on a turbine, a diminution in the effective delivery from 0.79 to

0.24, when the free opening under the wheel descended from its greatest elevation to about $\frac{1}{5}$ of this height. The inconvenience is so much the greater as the diminution in the effective delivery corresponds to that of the volume of water expended, which tends to make the dynamic effect of the machine very irregular. To remedy this, M. Fourneyron has proposed to subdivide the height of the turbine into several stages, by means of two or three annular horizontal plates, like the plates S R M N, U T P Q, the distance between which is divided into three or four equal parts. Supposing, for example, that there are three stages, we see that there will be no sudden change of section when the cylindrical sluice is raised $\frac{1}{3}$, $\frac{2}{3}$, or the whole height of the turbine; in all cases the phenomenon of sudden expansion will only affect a portion of the liquid vein. But, on the other hand, the construction of the machinery is complicated, and the friction of the water against the solid walls is increased. M. Fourneyron has again proposed using only the two plates S R M N, U T P Q; the lower one would carry the floats only, and the upper one be pierced with grooves which would allow it to settle freely between the floats, under the action of its weight alone. This upper plate would bear on a flange placed at the bottom of the cylindrical sluice on the outside. When the sluice is lowered, the plate S R M N is lowered with it, and the height of the turbine would be always equal to the height to which the sluice is raised. The movable plate carried around by the turbine would, while turning, rub against the flange which serves it as a support; but, the pressure between the two being slight, this would not produce an increase of resistance worthy of mention.

M. Fontaine, to regulate the expenditure of his turbines, uses a series of valves similar to that represented in Fig. 17. A B is a directing partition, B C a float of the turbine, D a valve which can be sunk to a great or less depth into the space between A B and the next directing partition to the left. In this way we can narrow as much as we choose the free passage into this interval, and as the same effect is produced on all in the same way, it is plain that we have the means of reducing the volume of water expended as much as circumstances may require. The vertical motion of translation is given simultaneously to all the rods E F by means similar to those used by M. Fourneyron: these rods are all connected with a metallic ring, at three points of which are attached vertical screws, furnished with nuts which can be turned only in their beds. These three have each a cogged wheel securely attached, the wheels being all of the same size, surrounded by an endless chain, which causes them to turn equally and at the same time. It is then necessary to turn one of the three wheels by means of a winch and pinion, in order that the whole system of valves may take up a vertical motion. The partial closing of the in-leading canals here, as in Fourneyron's turbine, is not without inconveniences, for the sudden change of section in these canals still gives rise to a loss of head; however, this loss is found to be diminished to a considerable extent.

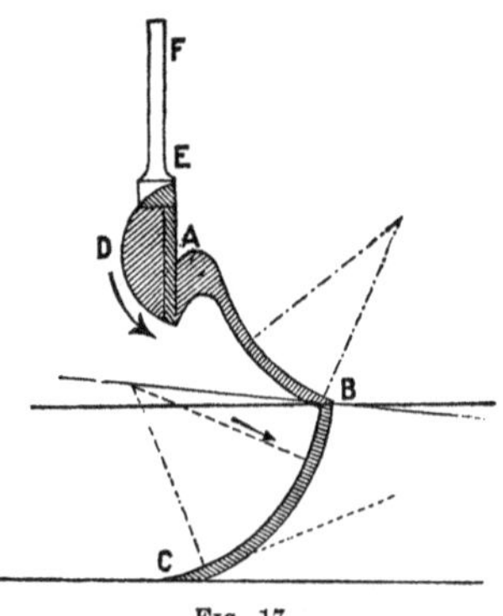

Fig. 17.

For the valves M. Kœcklin has substituted clack-valves revolving on a hinge, so as to fit exactly over the entrance to the distributing canals, when the closing is complete. The arrangement of the Kœcklin turbine also allows the expenditure of water to be regulated by making use of the sluice V (Fig.

15), which can close the communication between the well placed below the turbine and the tail race. But experience shows that a greater portion of the head is lost in this way than by using the valves.

The inconveniences attendant upon the partial closing of the distributing channels are so great, when considered from the point of view of economy of the motive power, that all possible means for remedying them have been sought for. We have already mentioned two devised by M. Fourneyron. M. Charles Callon, civil engineer, and a constructor of reputation, has proposed another way, which consists in making all the partial sluices, which close the channels in question, independent of each other; to diminish the expenditure, a certain number of these sluices can be closed completely, leaving the remainder entirely opened. But as the channels formed by the floats of the turbine pass alternately before the opened and closed orifices, there is still a cause of unsteadiness and trouble in the motion.

M. Callon's idea has been reproduced under another form by M. Fontaine. The orifices of entrance of the distributing channels occupy a horizontal surface comprised between two circles concentric with the axis of the turbine, M. Fontaine arranges two rollers, of the shape of a truncated cone, which can roll over this annular surface. The two rollers are mounted on the same horizontal spindle which forms a collar surrounding the axle of rotation. When they move in one direction, each one of them unrolls a band of leather, which has one end fastened to the roller and the other to the plane of the orifices of entrance: some of the orifices are thus entirely closed while the others remain wide open. When the truncated cones move in the contrary direction, they roll up the two leathern bands and uncover the orifices. M. Fontaine has also imitated the turbines of several stories of M. Fourneyron, in proposing to divide the turbines into several zones by means of surfaces of revolution about the axis of the system; each of these zones could be separately obstructed.

19. *Hydropneumatic turbine of Girard and Callon.*—The problem of regulating the expenditure, without too great loss, appears to have been solved in the best manner in a kind of turbine called by its inventors, MM. Girard and Callon, the *hydropneumatic* turbine. Their system consists essentially in surrounding Fourneyron's turbine with a sheet-iron bell, the lower plane of which is nearly at the height of the points at which the water leaves the floats. In this bell, by means of a small pump set in motion by the machine itself, the air is compressed, which gradually forces the water entirely out of the bell; then, if we suppose the cylindrical sluice to be partially raised, the liquid vein which escapes below has a depth less than the distance between the two plates of the turbine; but no sudden change in the section of liquid results from this, because the turbine moves in compressed air, and it is not covered by the water in the tail race. The water flowing into the turbine has a depth which, at the beginning of the floats, is equal to the height to which the sluice is raised; the upper plate is no longer wet, and as it only serves to hold the floats, it can be hollowed out, in order to make sure of a free circulation of air above the liquid vein. Thus the principal cause of the loss of head, due to the partial raising of the sluice, is found to be suppressed, and we should succeed in obtaining a very slightly varying effective delivery.

The calculations to be employed for the hydropneumatic turbine may be regarded as a particular case of those in No. 15. Preserving the same notation, we must consider b' as an unknown quantity, and at the same time suppose

$$p = p' = p_a + \Pi h;$$

in fact, h represents the depth to which the turbine is immersed below the level of the tail race, and $p_a + \Pi h$ is properly the pressure of the air in the bell. Thus, from this value of p equations (19) and (20) show first that $k = 1$, and consequently (No. 16) that we have $2\beta + \theta = 180°$, the only condition to be satisfied by the dimensions of the machine. It gives $\beta + \theta = 180° - \beta$; equation (17) then becomes

$$b' r'^2 \sin \gamma = b r^2 \sin \theta = b r^2 \sin 2\beta;$$

whence we can deduce b' in terms of the height b to which the sluice is raised, for a turbine moving with the maximum effective delivery. In virtue of the preceding relation, equations (13), (16), and (21) take the form

$$u^2 = g\,\mathrm{H} \frac{\tan \beta}{\sin \gamma} = \frac{g\,\mathrm{H}}{2 \cos^2 \beta},$$

$$\mathrm{Q} = 2\pi b r \sin \beta \sqrt{2 g \mathrm{H}},$$

$$\mu = 1 - \frac{r'^2}{2 r^2 \cos^2 \beta} (1 - \cos \gamma);$$

these formulas, which can very readily be demonstrated directly, give: 1st, the angular velocity $\frac{u}{r}$ of the turbine, which answers to the maximum effective delivery; 2d, its expenditure Q, and its effective delivery μ under the same condition. If we had to set up a turbine for a given head and expenditure, the equation of which Q is the first member, together with the inequality (24) (No. 16), would give us the means for finding r and $b \sin \beta$, and consequently b, after β has been chosen. As to γ, it should always be taken between 20 and 30 degrees; θ should be equal to $180° - 2\beta$; finally, r' should be taken as small as possible (on account of the expression for μ), but not, however, so as to run any risk of making the floats too short.

The method of MM. Girard and Callon could be equally well adapted to Fontaine's turbine.

20. *Some practical views on the subject of turbines.*—The directing partitions and floats are generally made of sheet iron; they are fastened to the surfaces that are to support them either by angle irons, or by setting them in cast-iron grooves on these surfaces. They should be sufficient in number to give to the velocity of the water their own direction. The distance of any two consecutive floats or partitions apart should not be, at any point, more than $0^m.06$ to $0^m.08$, measured along the normal to the surfaces, and generally it is made less. However, it must not be made too small, for then the friction of the water against the solid sides would be too great.

As the floats in Fourneyron's turbine are placed further from the axis than the partitions, that is to say distributed over a greater circumference, their number is from one-third to one-half greater than that of the partitions, in order to have everywhere a suitable distance apart.

Excepting the condition of cutting the planes of the orifices under a determinate angle, the curvature of the floats and partitions is almost a matter of indifference. However, as too great a curvature, or a sudden change in the curve, may prevent the threads from following the sides, and thus produce losses of head, we must avoid these two faults. It would be well to have the radius of curvature at least three or four times the distance apart measured along the normal.

Turbines can be used for all heads and every expenditure. For example, some are mentioned the head being only from $0^m.30$ to $0^m.40$, whilst, in the Black Forest, there is a turbine set up by M. Fourneyron which has a fall of 108 metres. The expenditure may be great even with quite small dimensions. In one of the examples given in No. 17, we have seen that a turbine of $0^m.60$ external radius and $0^m.09$ in height, expended without loss $1^{mc}.50$, or 1500 litres per second. Some turbines are constructed whose expenditure reaches as high as 4 cubic metres, or 4000 litres, per second, and, if need be, more could be expended.

Under ordinary circumstances, turbines move with sufficient rapidity, and thus allow the gearing for transmission to be economized.

For each turbine organized with fixed conditions, the theory of No. 15 gives a certain velocity to be imparted to it, in order to obtain the maximum useful effect. But if, in reality, a velocity different from this be given to it, and deviate from 25 per cent. more or less, experience shows that the effective delivery does not change much, a very important property for many manufacturing purposes, in which, in spite of the variations that take place in the expenditure of the head of water, it is important to have the machine always move with a very nearly constant velocity.

These motors have then a very decided advantage over wheels with a horizontal axis. Unfortunately their proportional effective delivery is not always constant, even approximately, for heads with a very variable expenditure; their construction and repairs can only be intrusted to very skilful mechanics, and consequently are quite expensive; while, on the other hand, overshot and breast wheels can be very economically constructed, and at the same time give an effective delivery at least equal to, if not greater than, that of turbines. These wheels will frequently, on this account, be preferred when the expenditure and head of water are favorable to their construction.

REACTION WHEELS.

---

21. *Reaction wheels.*—Let us conceive of one of Fourneyron's turbines deprived of its directing partitions, care being taken to prolong the floats to a slight distance from the axis of rotation; let us suppose that the water comes to the floats through a pipe concentric with the axle, having for a radius precisely the free distance that we have just mentioned. Furthermore, the expenditure of water in this pipe will be considered as very little, in order that the absolute velocity of the liquid in it may not be sensible. We shall thus have the idea of reaction wheels.

To give their theory, we will consider the point of entrance of water into the wheel as being on the axis of rotation itself; at this point the velocity of the wheel being nothing, as well as the absolute velocity of the water, it will be the same for the relative velocity of this last. In the calculations of No. 15, we will have to make $v = o$, $u = o$, $w = o$; equations (1), (2), and (3) then become

$$o = \mathrm{H} + h + \frac{p_a - p}{\Pi}$$

$$w'^2 = 2g\left(h' + \frac{p - p'}{\Pi}\right) + u'^2$$

$$\frac{p_a}{\Pi} + h + h' = \frac{p'}{\Pi}$$

whence we find

$$w'^2 = 2g\,\mathrm{H} + u'^2.$$

Neglecting friction, the only loss of head is as yet the height $\frac{v'^2}{2g}$ due to the absolute velocity of exit; we calculate its value by means of equation (6), which, combined with the preceding, will give

$$v'^2 = 2\,u'^2 + 2\,g\,\mathrm{H} - 2\,u' \cos\gamma \sqrt{2\,g\,\mathrm{H} + u'^2}.$$

The effective delivery would then be

$$\mu = \frac{\mathrm{H} - \frac{v'^2}{2\,g}}{\mathrm{H}} = 1 - \frac{v'^2}{2\,g\,\mathrm{H}} = -\frac{u'^2}{g\,\mathrm{H}} + \cos\gamma \frac{u'}{\sqrt{g\,\mathrm{H}}}$$

$$\times \sqrt{2 + \frac{u'^2}{g\,\mathrm{H}}},$$

or, placing $\frac{u'}{\sqrt{g\,\mathrm{H}}} = x$

$$\mu = -x^2 + x \cos\gamma \sqrt{2 + x^2}.$$

We can consider μ as a function of x, and seek its maximum when x varies. To this end we will take off the radical sign by writing

$$(\mu + x^2)^2 = x^2 \cos^2\gamma\,(2 + x^2),$$

or by reducing

$$x^4 \sin^2\gamma - 2\,x^2\,(\cos^2\gamma - \mu) + \mu^2 = 0.$$

If we deduced from this x^2 in terms of μ, the roots would become, from their nature, real; consequently, we have the condition

$$(\cos^2\gamma - \mu)^2 - \mu^2 \sin^2\gamma > 0,$$

or developing and reducing

$$\cos^4\gamma - 2\,\mu \cos^2\gamma + \mu^2 \cos^2\gamma > 0.$$

If we suppress the positive factor $\cos^2\gamma$, we find

$$\cos^2\gamma - 2\mu + \mu^2 > 0,$$

or

$$(1 - \mu)^2 - \sin^2\gamma > 0,$$

and, observing that $1 - \mu$ is necessarily positive, as well as $\sin \gamma$,

$$1 - \mu > \sin \gamma,$$

$$\mu < 1 - \sin \gamma.$$

The limit of the effective delivery that we can reach is then

$$\mu_1 = 1 - \sin \gamma.$$

The corresponding value for x is easily obtained by the relation between x and μ; if we take the equation deprived of radicals, and make in it $x = x_1$ and $\mu = 1 - \sin \gamma$, it becomes

$$x_1^4 \sin^2 \gamma - 2 x_1^2 \sin \gamma (1 - \sin \gamma) + (1 - \sin \gamma)^2 = 0,$$

or, more simply, by extracting the square root

$$x_1^2 \sin \gamma - (1 - \sin \gamma) = 0,$$

whence

$$x_1 = \sqrt{\frac{1 - \sin \gamma}{\sin \gamma}}.$$

If we supposed $\gamma = o$, we would find $\mu_1 = 1$; but x_1, and consequently u', would become infinite. Strictly speaking, the value $\gamma = o$ might be realized; it would only be necessary that the channels that make up the wheel should be arranged no longer side by side without empty spaces, as in turbines, but according to the annexed sketch (Fig. 18). We should have a certain number of curved plates, as A B, meeting the circumference O B at B, where they end, and communicating at A with the supply-pipe, to which they are permanently attached. The supply-pipe would then form the axle of rotation. But, in adopting this arrangement, u' could

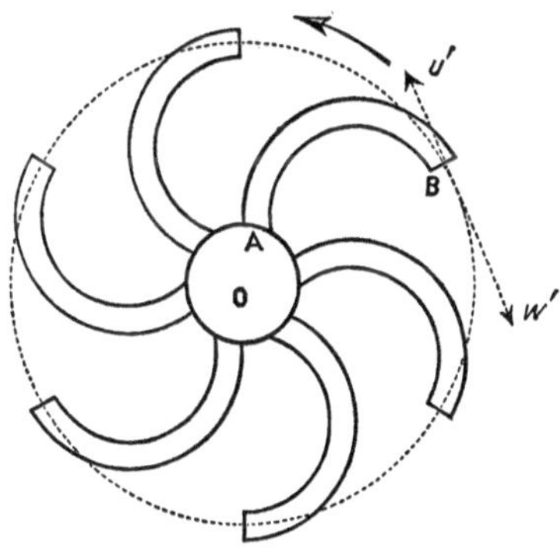

FIG. 18.

not become infinite, nor μ, consequently, reach unity: we only see that it would be necessary to make the wheel turn very rapidly. Moreover, it is difficult to expend a large volume of water without giving a considerable diameter to the central pipe A O, which would make the hypothesis, that u, v, and w were equal to zero, inadmissible; there are equally great difficulties in regulating the expense according to the required circumstances. It is undoubtedly on these accounts that this kind of wheel is but little used.

Preserving the arrangement of the floats of Fourneyron's turbine, which gives a series of contiguous channels, we can no longer make $\gamma = 0$, and then the limit of the theoretical effective delivery decreases quite rapidly as γ increases; so that, for $\gamma = 15°$, $1 - \sin \gamma$ would no longer be greater than 0.741. On the other hand, as we offer the water a wider outlet, we should perhaps lose less in friction, and the theoretical effective delivery would differ less from the real.

PUMPS.

---

§ IV. Of a few Machines for Raising Water.

22. *Pumps.*—The arrangement and shape of the parts of pumps are of infinite variety, according to the notions of the constructor. A special treatise would be necessary merely to describe the principal kinds. We will suppose, then, that the reader has seen a summary description of these machines, and confine ourselves to general ideas.

(*a.*) *Effort necessary to make the piston move.*—Two cases must be distinguished: pumps of single stroke, and pumps of double stroke. In the former case, the piston only draws up the water into the pump, or else only drives out the water previously drawn up, by forcing it up through the delivery pipe, when it moves in a determined direction; in the second, these effects are produced simultaneously, whatever be the direction of the stroke. Let us first take the single stroke; let

h be its height;

Ω the section of the piston;

p_a the atmospheric pressure;

Π the weight of a cubic metre of water.

The side of the piston in contact with the column of water drawn up would support, supposing that it remained in equilibrio, a pressure equal to $\Omega\,(p_a - \Pi\,h)$, whilst the other side, generally in direct communication with the atmosphere, would sustain a pressure in a contrary direction equal to $p_a\,\Omega$; the difference

$\Pi \Omega h$ would be the resultant pressure on the piston. If, on the contrary, there be but a single stroke forcing to a height h', we will find, in like manner, that the piston supports, exclusively of its motion, a resultant pressure $\Pi \Omega h'$. Finally, if the pump were one of double stroke, these two resultants would be superposed, and the value for the total pressure would be $\Pi \Omega (h + h')$ or $\Pi \Omega H$, H being the height included between the level of the basin from which the water is drawn, and that of the basin receiving it. In a certain class of single-stroke pumps, the same superposition of resultant pressure on the two faces of the piston takes place during the motion in one direction, and these resultants are in equilibrio when motion occurs in the opposite direction: this is the case in the kind of pump called the *lifting-pump*. If the piston have no horizontal motion, we must, besides, consider its weight as well as that of its rod, the component of which, parallel to the axis of the pump, would be added to or subtracted from the preceding expressions, according to circumstances. The friction of the piston against the barrel of the pump must also be added, as well as that of the rod against the packing-box, if there be any, through which it passes.

But these expressions only give the value of the force capable of maintaining the piston, as well as the water drawn or forced up, in equilibrio in a given position. When motion takes place, the effort brought to bear on the piston may be very different from this force. In the first place, the water does not begin to move in the pipes, and does not pass through the narrow openings of the valves without experiencing losses of head which are to be added to the heights h and h'. For example, in the case of the sucking pump, if there be a loss of head ζ in the length of the column drawn up, the pressure $\Omega (p_a - \Pi h)$ will be reduced to $\Omega (p_a - \Pi (h + \zeta))$, and the resultant $\Pi \Omega h$ would become $\Pi \Omega (h + \zeta)$. In like manner, if we consider a single forcing stroke, and that there is in the entire column forced up a loss of head ζ', the piézometric level in this column, at the point at which it touches the piston, would be increased by ζ', and the resultant pressure would be $\Pi \Omega (h' + \zeta')$. If the pump be one of double stroke, the expression $\Pi \Omega H$ should in like manner be replaced by $\Pi \Omega (H + \zeta + \zeta')$. Besides the heights ζ and ζ', others must be added, if the mass of water set in motion is not uniformly displaced. Let P be the weight of the piston and its rod, j its acceleration, P′ the weight of the water set in motion and which fills the pipes through which it is drawn up or forced out: the weight P′ being supposed to move with a mean acceleration j', we see that an additional force $\frac{1}{g}(Pj + P'j')$ would be necessary to overcome the inertia of the water and piston. This additional force, at one time a pressure, at another a resistance, may produce considerable variations in the total force to be applied to the piston, which is always inconvenient: since we must, in the first place, determine the dimensions of the pieces, not according to the mean, but according to the maximum effort sustained, which generally produces a clumsy and expensive machine; in the second, it is seldom that great variations in resistance do not give rise indirectly to some loss of motive power. We diminish the value of the term $\frac{Pj}{g}$ by means of weights acting as a counterpoise to the piston, if P is large; j' and consequently $\frac{P'j'}{g}$ are also diminished, in case it is found to be worth the trouble, by means of an air-chamber, placed at the entrance to the ascent pipe, which makes the motion of a great part of the weight of the column forced up uniform.

and consequently suppresses or diminishes to a very great extent the corresponding force of inertia.

Another means of obtaining approximate uniformity of motion in the ascent pipe consists in making it answer for the delivery of several pumps working together, in such a way that their total delivery in a series of equal times shall vary but little: it is accomplished in this way. Let O (Fig. 19) be the axis of rotation of an arbor that receives from a motor a motion rendered regular by a fly-wheel, and consequently nearly uniform. This arbor carries two arms, O B, O B′, making a right angle with each other; to each of them is attached a connecting rod, fastened at its other end to a piston running between guides, and which belongs to a double-stroke pump. As an example, we will suppose the arbor O horizontal, the piston rods vertical and their prolongations intersecting the axis of rotation; the connecting rods will generally be of the same length, about five or six times as long as the arm O B. It follows from this that the obliquity of the connecting rods with the vertical being always small, the velocities v and v' of the pistons are practically those of the projections of B and B′ on the vertical $B_0 B_1$; calling ω the angular velocity of the arbor, b the length O B, x the angle formed by O B with the vertical $B_0 B_1$, we will then have

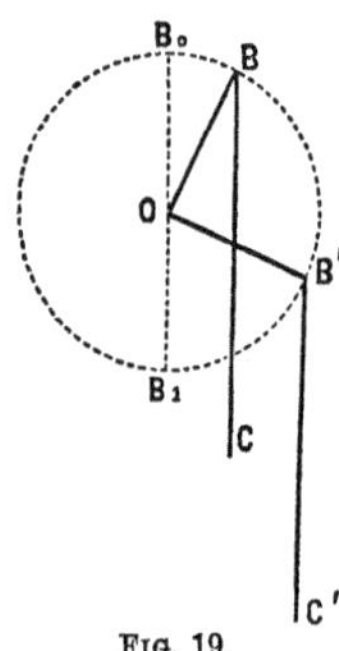

FIG. 19.

$$v = \omega b \sin x, v' = \omega b \sin\left(\frac{\pi}{2} + x\right) = \omega b \cos x.$$

Let Ω again be the common cross-section of the two pistons, and θ a very short interval of time; exclusively of the losses by the play of the two mechanisms, the volume of water furnished during the time θ to a common ascent pipe, by the two pumps together, will be the arithmetical sum of the volumes generated by the two pistons, viz., $\Omega\,(v + v')\,\theta$ or $\Omega\,\omega\,b\,\theta\,(\sin x + \cos x)$, a formula in which the sine and cosine should have their absolute values given, since it is a question of an arithmetical sum, and therefore the velocities are essentially positive. Consequently it is sufficient, in order to obtain the maximum, the minimum, and the mean of the variable quantity $\sin x + \cos x$, to suppose x included between o and $\frac{\pi}{2}$. Now we find between these limits

Two minima equal to 1 for $x = o$ and $x = \frac{\pi}{2}$

One maximum equal to 1.414, for $x = \frac{\pi}{4}$;

$$\text{The mean value equal to } \frac{\int_0^{\frac{\pi}{2}} (\sin x + \cos x)\,d\,x}{\int_0^{\frac{\pi}{2}} d\,x} = \frac{4}{\pi} = 1.272.$$

There would thus be between the minimum and the mean a relative difference of $\frac{0.272}{1.272}$, or about 0.214; whilst, with a single pump, the elementary delivery, proportional to $\sin x$, would vary between 0 and 1, and would have $\frac{2}{\pi}$ or 0.637 for a mean value, which would produce a much greater relative difference between the minimum and the mean.

We obtain a still more satisfactory result when we use three arms making angles of 120° with each other. The elementary delivery of the three pumps together is then proportional to

$$\sin x + \sin\left(x + \frac{2\,\pi}{3}\right) + \sin\left(x + \frac{4\,\pi}{3}\right),$$

each sine to be always taken as positive, whatever may be x.

We readily see, moreover, that the arithmetical sum of the three sines will not change by increasing the arc by 60 degrees, so that it is sufficient to make x vary between 0 and $\frac{\pi}{3}$. Within these limits, the first two sines are positive and the third negative; hence, the sum of the absolute values will be expressed by

$$\sin x + \sin\left(x + \frac{2\pi}{3}\right) - \sin\left(x + \frac{4\pi}{3}\right),$$

or else, by developing and observing that the arcs $\frac{2\pi}{3}$ and $\frac{4\pi}{3}$ together make an entire circumference,

$$\sin x + 2 \cos x \sin \frac{2\pi}{3}$$

or finally

$$\sin x + \sqrt{3} \cos x.$$

The minima of this quantity correspond to $x = 0$ and $x = \frac{\pi}{3}$, and have for their value $\sqrt{3}$ or 1.732; the maximum, corresponding to $x = \frac{\pi}{6}$, is $\frac{1}{2} + \frac{1}{2}\sqrt{3}.\sqrt{3}$, that is 2; the mean $\frac{\int_0^{\frac{\pi}{3}} (\sin x + \sqrt{3} \cos x)\, d\, x}{\int_0^{\frac{\pi}{3}} d\, x}$ becomes $\frac{6}{\pi}$ or 1.910. The relative difference between the minimum and the mean is consequently lowered to $\frac{1.910 - 1.732}{1.910}$ or to about 0.093.

There is also a great deal of regularity in the elementary delivery of the three pumps united as above, when they are supposed to be of a single stroke. Let us admit, for example, that each piston only forces up water when its crank O B descends from B_0 to B_1; the sum of the elementary deliveries will still be proportional to the expression

$$\sin x + \sin\left(x + \frac{2\pi}{3}\right) + \sin\left(x + \frac{4\pi}{3}\right);$$

but the pumps being only of single stroke, instead of changing the sign of the negative sines, we must suppress them altogether. This granted, let us first increase x from 0 to $\frac{\pi}{3}$: x and $x + \frac{2\pi}{3}$ will be less than the semi-circumference, and $x + \frac{4\pi}{3}$ will be included between π and 2π. We shall then only have to keep within these limits the sum $\sin x + \sin\left(x + \frac{2\pi}{3}\right)$, which can be put under the form

$$2 \sin\left(x \quad \frac{\pi}{3}\right) \cos\frac{\pi}{3}, \text{ or } \sin\left(x + \frac{\pi}{3}\right),$$

since $\cos\frac{\pi}{3} = \frac{1}{2}$; this sum, equal to $\sin\frac{\pi}{3}$, or 0.866 for $x = 0$, becomes a maximum and equal to 1 for $x = \frac{\pi}{6}$, then it decreases to 0.866 when x passes from $\frac{\pi}{6}$ to $\frac{\pi}{3}$. In the second place, if we take the values of x between $\frac{\pi}{3}$ and $\frac{2\pi}{3}$, the sines of $x + \frac{2\pi}{3}$ and $x + \frac{4\pi}{3}$ are both negative, so that the sin x alone can be preserved, which has still a minimum value 0.866, corresponding to two limits, and 1 for a maximum equidistant from these limits. Moreover, it is useless to consider the values of x greater than $\frac{2\pi}{3}$, for a rotation of 120 degrees not producing any change of figure in the combination of the mechanism, we would again find the same sines. We see, then, that the elementary delivery of the three pumps working together varies as certain numbers which are always comprised between 0.866 and 1, and consequently that it is sufficiently regular: the minimum and maximum are respectively half of what they were in the case of three pumps of double stroke.

We have heretofore supposed that the two cranks at right angles to each other, or the three cranks following each other at distances of 120 degrees, are fixed to the same arbor; it is plain that they can be attached to different arbors, provided that they all have the same angular velocity, the cranks being of equal length; or, more generally, provided that the centre of the joint of each of them with the corresponding connecting rod, in the three systems, has the same velocity of rotation about its arbor.

It is always well, as has been already said, to avoid great variations in the force that is to be transmitted to the piston of a pump; this becomes almost indispensable when it is moved by means of horses. An essential condition for employing the work of horses to advantage is that their pace and the exertion they have to make shall be as steady as possible; it would be difficult to accomplish this with one single stroke pump forcing up a long column of water, the piston of which should receive its motion from the arbor of the gearing-wheel by a system of a crank and connecting rod, and we could only succeed by employing fly-wheels, more or less clumsy. It would generally be better to regulate the resistance by means now to be mentioned.

(*b*) *Work to be transmitted to the piston.*—If we knew exactly, in each position of the piston, the force to be applied to give it motion, it would be easy to deduce the amount of work to be transmitted to it. But this force cannot be exactly determined; thus, the determination of friction of the piston against the pump barrel, or against the packing that it passes through (if it be a plunger), is necessarily uncertain, because it depends on the skill of the constructor; we can say the same for the losses of head encountered in the sucking and forcing pipes, on account of the defect of permanence and uniformity in the motion. However, when the forcing pipe is very long, we have seen that it is well so to arrange matters that the motion in it shall be uniform, and then we can calculate with sufficient exactness the entire head ζ'' between the two extremities of this pipe. This granted, let us admit first, that we are considering a double-stroke pump: the resultant pressure exerted on the piston being expressed by $\Pi\, \Omega\, (H + \zeta + \zeta')$, the force to be transmitted to it will be represented by $\Pi\, \Omega\, (H + \zeta'') + F$. We will include in the additive term F the friction against the barrel of the pump and the packing, the excess of $\zeta + \zeta'$ over ζ'', and lastly inertia. The total work of this force, in a linear distance l, divided into elements $d\,x$, will be

$$\Pi\, \Omega\, l\, (H + \zeta'') + \int_0^l F dx.$$

Moreover, $\Omega\, l$ represents very nearly the volume of water raised by one stroke of the piston; if then we wish the work expended in raising each cubic yard of water to the height H, we will have to calculate the quantity

$$(H + \zeta'') + \frac{1}{\Omega\, l} \int_0^l F dx.$$

In practice, on account of the difficulty of determining exactly the integral $\int_0^l F dx$, we simply multiply the term $\Pi\, (H + \zeta'')$ by a co-efficient n such as 1.10 or 1.15 or 1.20, according to the greater or less perfectness of the machine.

If the pump were one of single stroke, we would obtain the

same expression for the work, by adding together two consecutive strokes of the piston.

When we wish to determine the amount of horse-power to be transmitted to the piston, we must also know the mean velocity u of the piston. We easily deduce the mean amount pumped up per second Ωu, if the pump is one of double stroke, or $\frac{\Omega u}{2}$ for a single-stroke pump; we multiply this amount by $n \Pi (H + \zeta'')$; dividing finally by 75, we have the horse-power sought.

(c) *Mean velocity of the piston; delivery of pumps.*—The mean velocity of the piston ought not to be very small, because, in order to pump up any considerable quantity of water, we would have to make the body of the pump and the lifting pipe very large, which would increase the cost of construction. But too great a velocity possesses also great inconveniences: firstly, we increase the losses of head in a very rapid proportion; then it may happen that the water furnished by the sucking pipe may not come up sufficiently fast to follow the piston, and that the pump barrel may not be filled at each stroke, which would cause a loss in the delivery, and a shock on the return of the piston in an opposite direction. On account of the difficulty of determining exactly the velocity of the water sucked up, we ordinarily adopt a mean velocity of the piston, about $0^{m}.20$ per second; we rarely go so high as $0^{m}.30$. It is clear that the limit may be increased in proportion as the piston moves to a less height above the basin from which the water is drawn, and the more care that has been taken to avoid losses of head in the sucking pipe.

The piston having a stroke the length of which is l and the cross section Ω, describes, during one of the periods employed in raising water to the upper basin, a volume Ωl; this volume would also be that of the water raised during the same time, if there were no leakage around the valves between the piston and the hollow cylinder in which it moves. On this account, the volume raised varies from 0.75 Ωl to Ωl; the co-efficient by which the volume described by the piston is to be affected varies with the care displayed in the construction and in keeping the pump in good order; under ordinary circumstances, we may take it from 0.90 to 0.92.

SPIRAL NORIA.

23. *Spiral Noria.*—This wheel consists essentially of a horizontal arbor O (Fig. 20), to which are fastened a certain num-

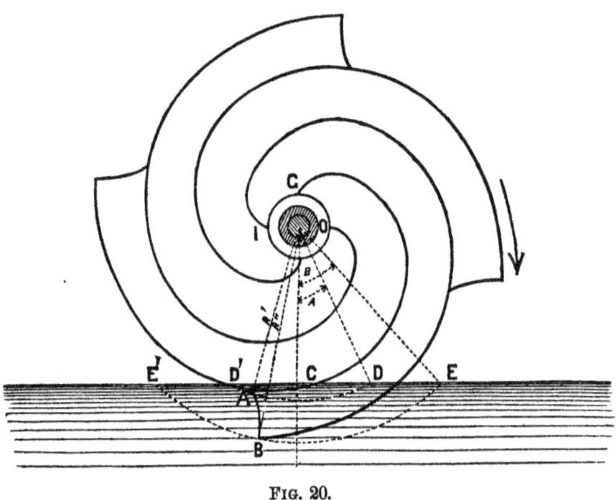

FIG. 20.

ber of cylindrical surfaces having their generatrices parallel to the axis; the right sections of these cylinders are involutes of a circle. The space between two consecutive cylinders thus forms a canal with a constant breadth, as well in the direction of the normal as perpendicularly to the plane of the figure. One of the canals, for example, will have its outer opening at A B and the other at I G. The entire system turns around the axis O, in the direction of the arrow-head; the centre O is above the basin whence the water is drawn, and the level of this basin covers the lower part of the wheel more or less.

During the time that the opening A B is wholly or partially

below the surface level of the basin, a certain quantity of water enters the canal A B I G, by the effect of the rotation; the rotation continuing, A B rises until finally it comes directly over I G; then the water taken in flows into I G through holes left open all around the arbor, and falls into a canal which leads it off to the basin that is to receive it.

We will propose two questions: 1st, a given spiral noria turns with a known angular velocity, and occupies a given position with respect to the lower basin; what will be the amount raised per second? 2d, what will be the work that the motor will have to transmit to it?

Let us call S the section A B projected on the plane passing through the axis O and the centre of A B; N the number of revolutions of the wheel per minute; n the number of involutes; H the height O C of the point O above the level of the water to be raised; r', r'' the distances of the points A and B from the axis of rotation. The point A will describe underneath the water an arc D A D′, of which we will designate the angle at the centre by 2α; in like manner the point B will describe the arc E B E′, corresponding to the angle at the centre 2β. First we shall have

$$\cos\alpha = \frac{H}{r'}, \cos\beta = \frac{H}{r''},$$

$$\text{arc D A D}' = 2r'\alpha = 2r' \cos^{-1}\frac{H}{r'},$$

$$\text{arc E B E}' = 2r''\beta = 2r'' \cos^{-1}\frac{H}{r''};$$

the arc described below the water by the centre of A B differing little from the mean $\frac{1}{2}$(D A D′ + E B E′) will then be expressed by

$$r' \cos^{-1}\frac{H}{r'} + r'' \cos^{-1}\frac{H}{r''} = L.$$

Now, the volume that entered at the opening A B is E D E′ D′, or the product of this arc by the perpendicular section S; then, since there are n canals that raise the same volume in each revolution of the wheel, the volume raised will be, per revolution, n L S; finally, the number of revolutions per second being $\frac{N}{60}$, the amount Q raised by the wheel in the same time will be expressed by $\frac{1}{60}$ N n L S, that is, we ought to have

$$Q = \frac{1}{60} N\, n\, S \left(r' \cos^{-1} \frac{H}{r'} + r'' \cos^{-1} \frac{H}{r''}\right).$$

But this calculation supposes that there enters, during each element of time $d\,t$ through the opening A B, a volume of water equal to that generated by A B in the same time; in this way the contraction which the liquid may experience on entering is not considered, nor is the motion communicated to the surrounding water, which, up to a certain point, may give way before the surface A B, instead of crossing it. For these reasons it would be well in practice to admit a certain reduction in the value of Q above given; we could effect it, for example, by a co-efficient which we will value, at a rough estimate, at 0.80, for want of exact experiments on this subject.

Here is an example for calculating Q. Let $N = 12$, $n = 4$, $r' = 2^{m}.50$, $r'' = 3^{m}.0$, $H = 2^{m}.0$, $S = 0^{sq.\,m}.17$. We shall have

$$\frac{H}{r'} = 0.8000, \quad \cos^{-1} \frac{H}{r'} = 0.410\, \frac{\pi}{2};$$

$$\frac{H}{r''} = 0.6667, \quad \cos^{-1} \frac{H}{r''} = 0.535\, \frac{\pi}{2};$$

$$r' \cos^{-1} \frac{H}{r'} = r'' \cos^{-1} \frac{H}{r''} = \frac{\pi}{2}\,(1.025 + 1.605) = 4.131;$$

whence we deduce

$$Q = 0^{mc}.562,$$

a number which would be reduced to $0^{mc}.45$ about, by multiplying by 0.80.

As to the motive work to be expended in raising a certain weight P of water, it is composed: 1st, of the work P H destined to overcome that of the weight; 2d, of the work of friction on the trunions and shoulders of the arbor O, which can be determined by means of known formulæ; 3d, of the work necessary to overcome the friction of the water against the solid walls with which it is in contact; this work being very slight, if the involutes form tolerably large channels; 4th, the work necessary to give to the water the absolute velocity with which it leaves the wheel. This last work will also be very slight, if we take care to make the wheel turn slowly; for the lowest point of any involute whatever being always on the vertical through I, we see that the water that has already entered the interior of the canal A B I G, and that which will still enter in the course of the same revolution, will only be completely emptied out after an entire revolution, reckoning from the position indicated by the figure. The water rises then with little absolute velocity into the machine, and consequently a small portion of the motive work is employed in giving to it an unproductive living force. But it must not be forgotten that this supposes slowness of revolution around the axis O.

To sum up, we will calculate the first two portions of the motive work, which are the most important, and in order to account approximately for the other two, we will multiply the sum of the calculated portions by a co-efficient a little greater than unity.

The first idea of the noria is very old, since Vitruvius speaks of a similar machine; it was Lafaye who, in 1717, proposed giving it the form we have described above. This machine seems susceptible of a very good delivery, and is well adapted to raising large volumes of water; but the height to which the water is raised, always less than the radius of the wheel, is necessarily very limited; besides, this wheel is heavy, and on this account hard to transport.

CENTRIFUGAL PUMP.

24. *Lifting turbines; centrifugal pump.*—The greater part of the machines which are used to turn the motive power of a head of water to account, can, with a few changes, be converted into machines for raising water, and the reverse. Thus, for example, if a breast-wheel, set in a water-course, receives a motion about its horizontal axis, by the action of any motor, so that the floats may ascend the circular flume, these floats will carry up with them the water from the tail race and throw it into the head race: we would then obtain, in principle, the lifting wheel. In like manner, let us take one of Fourneyron's turbines, and make the intervals between the directing partitions communicate directly with the tail race, and let the outer orifices of the turbine open into a compartment from which the ascent pipe leads; when a motion of rotation is impressed on the wheel, the water contained in the floats will be urged toward the exterior by the centrifugal force, and will reach the enclosed compartment with an excess of pressure which will cause it to ascend the pipe to a certain height, the greater as the rotation becomes more rapid. If the pipe is not too high, a delivery of water will take place at its end; and this, moreover, will be continuous, the water thrown out by the centrifugal force being incessantly replaced by that from the tail race, which tends to fill up the empty space between the partitions.

The theory of such a turbine, which we might call a *lifting turbine*, resembles very closely that of (No. 15). But as the machine there discussed has not as yet been set up or experimented upon, it need no longer be dwelt upon. We will pass to the study of a pump called the *centrifugal pump*, which belongs to the same class of machines, but which bears, however, a greater resemblance to reaction wheels.

A wheel composed of a series of cylindrical floats, such as B C (Fig. 21), assembled between two annular plates, is caused

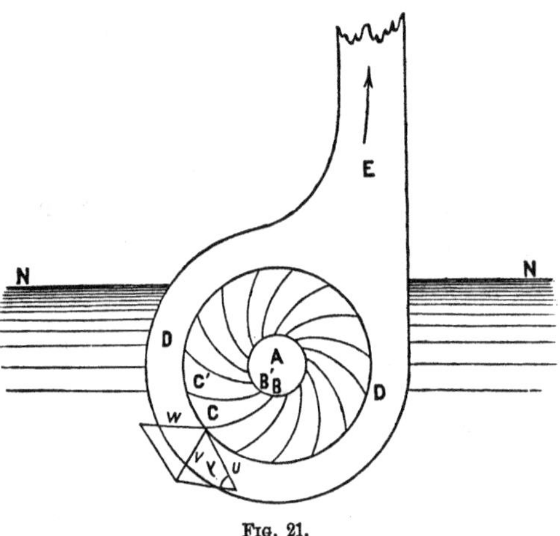

FIG. 21.

to turn around a horizontal axis projected at A. The water, from the basin to be emptied, comes freely within the circle A B, which limits the floats on the inside, either because the centre A is a little below the level N N of this bay, or by means of suction pipes. The motion of rotation impressed on this wheel drives the water from the canals B C, B′ C′, into the annular space D, where it acquires a pressure sufficient to drive it up the pipe E, the only means of escape open to it, and by which it reaches the upper basin. The angular velocity of the arbor A being known, as well as all the dimensions of the machine, and its position relatively to the basins of departure and arrival, we can find the amount of water pumped up per second, the motive work that it requires, and its delivery.

To show this, let us call

H the difference of level between the two basins;

h the depth of the centre A below the level of the lower;

r the exterior radius A C of the wheel;

b the distance apart of the two annular plates, which confine the floats;

ω the angular velocity of the arbor A;

v the absolute velocity of the water when it leaves the floats;

u the velocity $\omega\ r$ at the outer circumference of the wheel;

w the relative velocity of the water at the same point;

γ the acute angle formed by the velocities w and u—that is, the angle at which the floats cut the outer circumference;

p the pressure of the water at its point of entrance into the interval between the floats;

p' its pressure at the point of exit;

p_a the atmospheric pressure;

Π the weight of the cubic metre of water.

We will begin by simplifying the question a little by means of a few hypotheses. First, we will neglect the absolute velocity of the water in the ascent pipe and in the conduit which conveys it to the floats, which may be allowed if the cross sections of these conduits are sufficiently large relatively to the volume pumped out. The radius A B, however, should still be sufficiently small so that the velocity of rotation of the point B may be neglected; in other words, we will consider the introduction of the water into the wheel as taking place along the axis, without any velocity occasioned by the motion, and consequently without any relative velocity. Secondly, we will conduct our argument as though the water were displaced horizontally in its passage across the wheel; the height of this last will be supposed slight relatively to H, so that it can be left out of account. Besides, nothing in practice would prevent our assuming the arbor A as vertical; but this would be a matter of very little importance in the result.

This granted, the pressure varying according to the hydrostatic law from the lower basin to the point of entrance, and from the point of exit to the other basin, we will have

$$p = p_a + \Pi h,$$
$$p' = p_a + \Pi (H + h);$$

whence, by subtraction,

$$\frac{p' - p}{\Pi} = H.$$

Now, if we apply Bernouilli's theorem to the relative motion of a molecule following the curve B C, the fictitious gain of head will be expressed by $\frac{\omega^2 r^2}{2g}$ or $\frac{u^2}{2g}$, and we shall find

$$\frac{w^2}{2g} = \frac{p - p'}{\Pi} + \frac{u^2}{2g},$$

or else

$$w^2 = -2gH + u^2, \quad . \quad . \quad . \quad (1)$$

an equation giving w since u is known. This first result gives the means of calculating the amount Q pumped up in each second. In fact, the water leaving the floats cuts a cylindrical surface $2\pi b r$ at an angle γ and with the relative velocity w; then the total orifice of exit, measured perpendicularly to w, is $2\pi b r \sin\gamma$, and consequently

$$Q = 2\pi b r w \sin\gamma. \quad . \quad . \quad . \quad (2)$$

The motive work consumed per second in making the wheel turn includes first the work $\Pi Q H$; then the water reaching the annular space D with a velocity v, this is lost in useless disturbance; whence there results a molecular work $\Pi Q \frac{v^2}{2g}$.

Thus then, throwing out of account the other frictions, the work expended per second will be $\Pi\, Q\left(H + \frac{v^2}{2\,g}\right)$; and as the useful work is only $\Pi\, Q\, H$, the effective delivery μ will have for its value

$$\mu = \frac{H}{H + \frac{v^2}{2\,g}} = \frac{1}{1 + \frac{v^2}{2\,g\,H}}. \quad . \quad . \quad . \quad (3)$$

There remains to determine v; now v is the resultant of w and u, hence we have

$$v^2 = u^2 + w^2 + 2\,u\,w \cos\gamma,$$

or from eq. (1)

$$v^2 = -\,2\,g\,H + 2\,u^2 - 2\,u\cos\gamma\,\sqrt{-\,2\,g\,H + u^2}. \quad . \quad . \quad . \quad . \quad (4)$$

Equations (1), (2), (3), and (4) give the means of solving without difficulty the questions proposed.

Let us again see by what means we could obtain the greatest possible result of the motive power. Expression (3) for the effective delivery becomes, substituting for v its value, and making $\frac{u}{\sqrt{g\,H}} = x$.

$$\mu = \frac{1}{x^2 - x\cos\gamma\,\sqrt{x^2 - 2}};$$

we shall then have the maximum of μ, considered as a function of x, in seeking the minimum of the denominator, or, what amounts to the same, the minimum of $\frac{1}{\mu}$. We shall conduct this research as in (No. 21); we will write

$$x^2 - \frac{1}{\mu} = x\cos\gamma\,\sqrt{x^2 - 2}$$

or, by making the radical disappear and transposing,

$$x^4\sin^2\gamma - 2\,x^2\left(\frac{1}{\mu} - \cos^2\gamma\right) + \frac{1}{\mu^2} = 0.$$

Now μ can only receive values that, substituted in this biquadratic equation, will give x^2 real and positive; hence we have

$$\left(\frac{1}{\mu} - \cos^2 \gamma\right)^2 - \frac{1}{\mu^2} \sin^2 \gamma > 0,$$

or successively

$$\frac{1}{\mu^2} \cos^2 \gamma - \frac{2}{\mu} \cos^2 \gamma + \cos^4 \gamma > 0,$$

$$\frac{1}{\mu^2} - \frac{2}{\mu} + \cos^2 \gamma > 0,$$

$$\left(\frac{1}{\mu} - 1\right)^2 > \sin^2 \gamma.$$

As $\sin \gamma$ and $\frac{1}{\mu} - 1$ are positive quantities, we can extract the square root of both members and place

$$\frac{1}{\mu} - 1 > \sin \gamma, \text{ or } \frac{1}{\mu} > 1 + \sin \gamma;$$

the minimum of $\frac{1}{\mu}$ has then for its value $1 + \sin \gamma$, and the limit of the effective delivery μ_1 will be $\frac{1}{1 + \sin \gamma}$. The corresponding value x_1 of x is obtained from the above biquadratic equation, which gives

$$x_1^2 = \frac{\frac{1}{\mu_1} - \cos^2 \gamma}{\sin^2 \gamma} = \frac{1 + \sin \gamma - \cos^2 \gamma}{\sin^2 \gamma} = \frac{1 + \sin \gamma}{\sin \gamma}.$$

Thus the most favorable velocity u for the effective delivery is obtained from this equation

$$u^2 = g \text{H} \frac{1 + \sin \gamma}{\sin \gamma};$$

whence we get the angular velocity $\omega = \frac{u}{r}$, and the number of revolutions per minute $\text{N} = \frac{30 \omega}{\pi}$. The effective delivery being then $\frac{1}{1+\sin\gamma}$, we may be tempted, in order to increase it, to make γ very small; but we see that the velocities u and w would become very great, and we should thus lose a great deal in the friction of the water against the floats. Besides, we have, from equations (1) and (2),

$$Q = 2\pi b r \sin\gamma \sqrt{u^2 - 2g\text{H}} = 2\pi b r \sqrt{g\text{H}} \sin\gamma . \sqrt{x^2 - 2};$$

the amount Q′, which corresponds to the maximum effective delivery, will then be

$$Q' = 2\pi b r \sqrt{g\text{H}} . \sin\gamma . \sqrt{\frac{1+\sin\gamma}{\sin\gamma} - 2}$$

$$= 2\pi b r \sqrt{g\text{H}} \sqrt{\sin\gamma(1-\sin\gamma)}.$$

This amount reduces to zero at the same time as γ; when γ alone varies, Q′ becomes a maximum for $\sin\gamma = 1 - \sin\gamma$, or $\sin\gamma = \frac{1}{2}$ or $\gamma = 30°$; the effective delivery is then $\frac{1}{1+\frac{1}{2}}$ that is $\frac{2}{3}$. The value $\gamma = o$ is consequently inadmissible as reducing the amount to zero; but from this point of view it is not well to go beyond $\gamma = 30°$. On the other hand, this last value does not give a very high theoretical effective delivery; perhaps the best thing to do in practice would be to take γ between 15 and 20 degrees. For $\gamma = 15°$, for example, the effective delivery increases to $\frac{1}{1+0.2588} = 0.794$, and the product $\sqrt{\sin\gamma(1-\sin\gamma)}$ is decreased to 0.438, whereas it is 0.50 for $\gamma = 30°$; it is a diminution that could be compensated for by a slight increase of r or b.

Instead of arranging the wheel as represented in Fig. 21, we might adopt two separate canals, as in Fig. 19. In this case the expression for the amount would change, and the angle γ

might become zero; but, whereas u cannot increase to infinity, the value $\frac{1 + \sin \gamma}{\sin \gamma}$ would no longer be admissible for $\frac{u^2}{g\,\mathrm{H}}$, and we should have to depart more or less from the limit of the effective delivery. Furthermore, for an equal expenditure, we should probably lose more in friction.

AUTHORITIES ON WATER WHEELS.

Expériences sur les Roues Hydrauliques à aubes planes, et sur les Roues Hydrauliques à augets, by Morin.

Experiments made by the Committee of the Franklin Institute on Water Wheels, in 1829–30. See Journal of the Franklin Institute, 3*d Series*, Vol. I., pp. 149, 154, &c., and Vol. II., p. 2.

Experiments on Water Wheels, by Elwood Morris. See Jour. Frank. Ins., 3*d Series*, Vol. IV., p. 222.

Mémoire sur les Roues Hydrauliques à aubes courbes, mues par dessous, by Poncelet.

Expériences sur les Roues Hydrauliques à axe vertical, appelées Turbines, by Morin.

Experiments on the Turbines of Fourneyron, by Elwood Morris. See Jour. Frank. Ins., Dec. 1843, and Jour. Frank. Ins., 3*d Series*, Vol. IV., p. 303.

Lowell Hydraulic Experiments, 2d Ed., 1868, by Jas. B. Francis.

APPENDIX.

Comparative Table of French and United States Measures.

Pounds avoirdupois in a kilogramme	2.2
Inch in a millimetre	0.039
Inch in a centimetre	0.393
Inch in a decimetre	3.937
Feet in a metre	3.280
Yard in a metre	1.093
Square feet in a square metre	10.7643
Cubic inch in a cubic centimetre	0.061
Cubic feet in a cubic metre	35.316
Quart in a litre	1.0567

NOTE.—A cubic metre of distilled water weighs one thousand kilogrammes.

The litre contains one cubic decimetre of distilled water, and weighs one kilogramme.

The horse-power of the French is 75 kilogrammetres, equivalent to $542\frac{1}{2}$ foot-pounds per second; the English horse-power being 550 foot-pounds per second. The value of g is 9.81 metres; and that of the height of a column of water equivalent to the atmospheric pressure is taken at 10.33 metres.

Note A. *Art.* 1.

The formula given in this article is to be found, with the exception of variations in the notation, in all works of applied mechanics in which the subject of the theory of machines is discussed (see, for example, Moseley's Engineering and Architecture, Am. Ed., p. 146). The only term in it which is not generally found in other works is the one $(H - H_0) \Sigma\, m\, g$, which expresses the work expended in overcoming the weight of any part of the machine, when its centre of gravity is raised from one level to another, represented by the vertical height $(H - H_0)$, and the corresponding work by $(H - H_0) \Sigma\, m\, g$; as, for example, in the case of a wheel revolving on a horizontal axle, the axis of which does not coincide with its centre of gravity; or in that of a revolving crank; in both of which cases the work expended will be equal to the product of the weight $\Sigma\, m\, g$ raised, and the vertical height $(H - H_0)$ passed over by its centre of gravity. But, in all like cases, as in the descent of the centre of gravity from its highest to its lowest position, the same amount of work will be restored by the action of gravity, the total work expended will be zero for each revolution, and the term $(H - H_0) \Sigma\, m\, g$ will disappear from the formula.

Note B. *Art.* 2.

The equation $H - \frac{1}{g}(t_e + t_f) - \frac{U^2 - U_0^2}{2\,g} = o$ is the modified form of what is known as Bernouilli's theorem as applied to the case treated of in Art. 2.

This theorem, applied to the phenomena of the flow of a heavy homogeneous fluid, may be generally thus stated: *The increase of height due to the velocity is equal to the difference between the effective head and the loss of head.*

In this case H is the effective head; $-\frac{1}{g}(t_e + t_f)$ is the loss of head from the dynamical effect t_e imparted to the wheel by the action of the water, and the work t_f due to the various secondary resistances; and the term $-\frac{U^2 - U^2_0}{2\,g}$ the height due to the velocity.

Multiplying each member of the above equation by g, we obtain

$$g\,H - t_e - t_f = \frac{1}{2}\,U^2 - \frac{1}{2}\,U^2_0,$$

or, in other words, the modified expression of the general formula *Note* A, as applied to this case.

See Bresse. *Mécanique Appliquée. Vol.* 2, *Nos.* 12 *and* 15, *pp.* 23 *and* 30.

Note C. *Art.* 4.

In the equation

$$F = \frac{P}{g}\,(v - v') - \frac{1}{2}\,\Pi\,b\,(h'^2 - h^2),$$

which expresses the force applied horizontally at the centre of the submerged portion of the bucket, the second term of the second member $\frac{1}{2}\,\Pi\,b\,(h'^2 - h^2)$, represents the diminution of the force imparted to the wheel by the current, arising from the increase of depth of the water as it leaves the wheel, or by the back water; or, in other words, the difference of level between the point C, before the depth of the current is affected by the action of the wheel, and the point E, where the depth of the current has increased from the back water. This difference of level receives the name of a *surface fall.*

The relations existing between the two terms of the second member of the equation, leaving out of consideration the action of the wheel, may be established in the following manner.

Considering the portion of the current comprised between the two sections C B, E F (Fig. 3), at a short distance apart, between which the surface fall takes place, we can apply to the liquid system C B E F, comprised between these sections, the theorem of the quantities of motion projected on the axis of the current, which, in the present case, may be regarded as horizontal. Now, during a very short interval of time θ, the system C B E F will have changed its position to C′ B′ E′ F′, and, in virtue of the supposed permanency of the motion, each

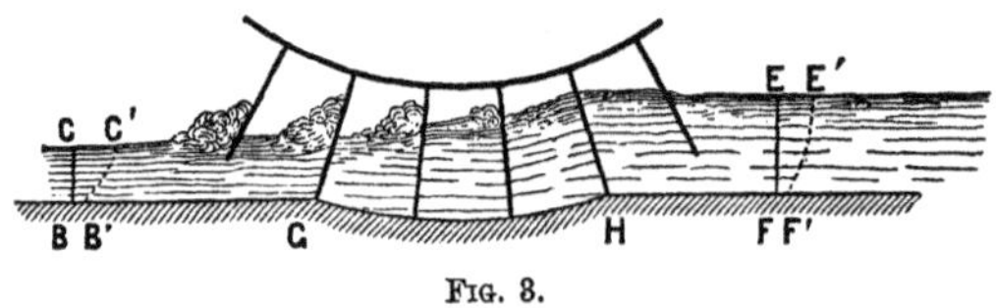

Fig. 8.

point of the intermediate portion C′ B′ E F will have equal masses moving with the same velocity at the beginning and ending of the time θ; the variation in the projected quantity of motion of the system C B E F, during the time θ, will therefore be equal to the quantity of motion of the portion included between the final sections E F, E′ F′, and that comprised between the initial sections C B, C′ B′.

To find these quantities of motion. Represent by ω' a superficial element of the section E F, and by v' the velocity of the fluid thread which flows through it; $v'\theta$ will then be the length of this thread for the time θ, between the sections E F and E′ F′; and $\omega' v' \theta$ will be the volume of the thread which has ω' for its base and $v'\theta$ for its length. Representing by Π the weight of a cubic metre of the liquid, $\frac{\Pi}{g}\omega' v' \theta$ will be the corresponding mass of this volume, and $\frac{\Pi}{g}\omega' v'^2 \theta$ its quantity of motion; and, designating by Σ the sum of all the elements ω',

$\frac{\Pi}{g} \Sigma \omega' v'^2 \theta$ will be the quantity of motion of the portion E F E′ F′ of the liquid comprised between the two final sections. In like manner, v being the velocity with which each thread flows through an element ω of the section B C, the quantity of motion of the portion of liquid between the sections B C and B′ C′ will be expressed by $\frac{\Pi}{g} \Sigma \omega v^2 \theta$. The increase, therefore, in the quantity of motion during the time θ will be expressed by

$$\frac{\Pi}{g} (\Sigma \omega' v'^2 \theta - \Sigma \omega v^2 \theta).$$

But as v and v' may be assumed as sensibly equal to the mean velocities of the current in sections E F, C B, then $\Sigma \omega' v'$ and $\Sigma \omega v$ will be the volumes corresponding to these velocities; and $\frac{\Pi}{g} \Sigma \omega' v'$ and $\frac{\Pi}{g} \Sigma \omega v$ the corresponding masses. But since, from the permanency of the motion $\omega' v' = \omega v$, the expression for the increase of the quantity of motion, for the time θ will therefore take the form

$$\frac{P \theta}{g} (v' - v),$$

in which P represents the weight of the water expended in each second, and $\frac{P}{g}$ its corresponding mass.

The expression here found is equal to the sum of the impulsions of the forces exterior to the liquid system considered during the time θ, also projected on the horizontal axis of the current. From the form given to the section of the race, which is rectangular, the direction of the axis of the current, which is assumed as horizontal, between the extreme sections, and the short distance between these sections, the only impulsions of the pressures upon the liquid system are those on the sections E F and C B. Representing, then, by b the breadth of the sections, by h' and h their respective depths, their respective areas will be expressed by $b\,h'$ and $b\,h$; the pressures on these areas will be $\Pi\, b\, h' \times \frac{1}{2} h'$ and $\Pi\, b\, h \times \frac{1}{2} h$; and for the respective projected sums of the impulsions of these pressures, during the time θ, we shall have $\frac{1}{2} \Pi\, b\, \theta\, h'^2$ and $\frac{1}{2}\, \Pi\, b\, \theta\, h^2$, since from the circumstances of the motion the pressures follow the hydrostatic law. The impulsion in the direction of the motion will therefore be expressed by $\frac{1}{2} \Pi\, b\, \theta\, (h^2 - h'^2)$, from which we obtain

$$\frac{P}{g}\, \theta\, (v' - v) = \frac{1}{2} \Pi\, b\, \theta\, (h^2 - h'^2).$$

to express the relation in question.

See Bresse. *Mécanique Appliquée. Vol.* 2, *No.* 83, *p.* 245.

Note D. *Art.* 9.

The term $C\left(1 - \frac{p}{2R}\right) - c$, which expresses the arc intercepted between two buckets, taken at the middle point of their depth, is obtained as follows:

R being the exterior radius of the wheel, corresponding to the arc C, the radius of the arc at the middle point of the bucket will be $R - \frac{1}{2} p$; calling x the arc corresponding to this radius, we have

$$R : R - \frac{1}{2} p :: C : x. \quad \therefore x = C\left(1 - \frac{p}{2R}\right);$$

and for the arc intercepted between the two buckets at their middle point the expression above.

Note E.

By the courteous permission of JAMES B. FRANCIS, Esq., granted through Gen. JOHN C. PALFREY, the following extracts were taken from the valuable work of Mr. Francis, under the title of "LOWELL HYDRAULIC EXPERIMENTS."

A VAST amount of ingenuity has been expended by intelligent millwrights on turbines; and it was said, several years since, that not less than three hundred patents relating to them had been granted by the United States Government. They continue, perhaps, as much as ever to be the subject of almost innumerable modifications. Within a few years there has been a manifest improvement in them, and there are now several varieties in use, in which the wheels themselves are of simple forms, and of single pieces of cast iron, giving a useful effect approaching sixty per cent. of the power expended.

In the journal of the Franklin Institute, Mr. Morris also published an account of a series of experiments, by himself, on two turbines constructed from his own designs, and then operating in the neighborhood of Philadelphia.

The experiments on one of these wheels indicate a useful effect of seventy-five per cent. of the power expended, a result as good as that claimed for the practical effect of the best overshot wheels, which had heretofore in this country been considered unapproachable in their economical use of water.

In the year 1844, Uriah A. Boyden, Esq., an eminent hydraulic engineer of Massachusetts, designed a turbine of about seventy-five horse power, for the picking-house of the Appleton Company's cotton-mills, at Lowell, in Massachusetts, in which wheel Mr. Boyden introduced several improvements of great value.

The performance of the Appleton Company's turbine was

carefully ascertained by Mr. Boyden, and its effective power, exclusive of that required to carry the wheel itself, a pair of bevel gears, and the horizontal shaft carrying the friction-pulley of a Prony dynamometer, was found to be seventy-eight per cent. of the power expended.

In the year 1846, Mr. Boyden superintended the construction of three turbines, of about one hundred and ninety horse-power each, for the same company. By the terms of the contract, Mr. Boyden's compensation depended on the performance of the turbines; and it was stipulated that two of them should be tested. In accordance with the contract, two of the turbines were tested, a very perfect apparatus being designed by Mr. Boyden for the purpose, consisting essentially of a Prony dynamometer to measure the useful effects, and a weir to gauge the quantity of water expended.

The observations were put into the hands of the author for computation, who found that the mean maximum effective power for the two turbines tested was eighty-eight per cent. of the power of the water expended.

According to the terms of the contract, this made the compensation for engineering services, and patent rights for these three wheels, amount to fifty-two hundred dollars, which sum was paid by the Appleton Company without objection.

These turbines have now been in operation about eight years, and their performance has been, in every respect, entirely satisfactory. The iron work for these wheels was constructed by Messrs. Gay & Silver, at their machine-shop at North Chelmsford, near Lowell; the workmanship was of the finest description, and of a delicacy and accuracy altogether unprecedented in constructions of this class.

These wheels, of course, contained Mr. Boyden's latest improvements, and it was evidently for his pecuniary interest that the wheels should be as perfect as possible, without much regard to cost. The principal points in which one of them differs from the constructions of Fourneyron are as follows:—

The wooden flume conducting the water immediately to the turbine is in the form of an inverted truncated cone, the water being introduced into the upper part of the cone, on one side of the axis of the cone (which coincides with the axis of the turbine), in such a manner that the water, as it descends in the cone, has a gradually increasing velocity and a spiral motion; the horizontal component of the spiral motion being in the direction of the motion of the wheel. This horizontal motion is derived from the necessary velocity with which the water enters the truncated cone; and the arrangement is such that, if perfectly proportioned, there would be no loss of power between the nearly still water in the principal penstock and the guides or leading curves near the wheel, except from the friction of the water against the walls of the passages. It is not to be supposed that the construction is so perfect as to avoid all loss, except from friction; but there is, without doubt, a distinct advantage in this arrangement over that which had been usually adopted, and where no attempt had been made to avoid sudden changes of direction and velocity.

The guides, or leading curves (Figs. A, B), *are not perpendicular, but a little inclined backwards from the motion of the wheel, so that the water, descending with a spiral motion, meets only the edges of the guides.* This leaning of the guides has also another valuable effect: when the regulating gate is raised only a small part of the height of the wheel, the guides do not completely fulfil their office of directing the water, the water entering the wheel more nearly in the direction of the radius than when the gate is fully raised; by leaning the guides it will be seen the ends of the guides near the wheel are inclined, the bottom part standing farther forward, and operating more efficiently in directing the water when the gate is partially raised, than if the guides were perpendicular.

In Fourneyron's constructions a garniture is attached to the regulating gate, and moves with it, for the purpose of diminishing the contraction. This, considered apart from the mechanical difficulties, is probably the best arrangement; to be perfect, however, theoretically, this garniture should be of different forms for different heights of gate; but this is evidently impracticable.

In the Appleton turbine the garniture is attached to the guides, the gate (at least the lower part of it) being a simple thin cylinder. By this arrangement the gate meets with much less obstruction to its motion than in the old arrangement, unless the parts are so loosely fitted as to be objectionable; and it is believed that the coefficient of effect, for a partial gate, is proportionally as good as under the old arrangement.

On the outside of the wheel is fitted an apparatus, named by Mr. Boyden the Diffuser. The object of this extremely interesting invention is to render useful a part of the power otherwise entirely lost, in consequence of the water leaving the wheel with a considerable velocity. It consists, essentially, of two stationary rings or discs, placed concentrically with the wheel, having an interior diameter a very little larger than the exterior diameter of the wheel; and an exterior diameter equal to about twice that of the wheel; the height between the discs at their interior circumference is a very little greater than that of the orifices in the exterior circumference of the wheel, and at the exterior circumference of the discs the height between them is about twice as great as at the interior circumference; the form of the surfaces connecting the interior and exterior circumferences of the discs is gently rounded, the first elements of the curves near the interior circumferences being nearly horizontal. There is consequently included between the two surfaces an aperture gradually enlarging from the exterior circumference of the wheel to the exterior circumference of the diffuser. When the regulating gate is raised to its full height, the section through which the water passes will be increased, by insensible degrees, in the proportion of one to four, and if the velocity is uniform in all parts of the diffuser at the same distance from the wheel, the velocity of the water will be diminished in the same proportion; or its velocity on leaving the diffuser will be one-fourth of that at its entrance. By the doctrine of living forces, the power of the water in passing through the diffuser must, therefore, be diminished to one-sixteenth of the power at its entrance. It is essential to the proper action of the diffuser that it should be entirely under water, and the power rendered useful by it is expended in diminishing the pressure against the water issuing from the exterior orifices of the wheel; and the effect produced is the same as if the available form under which the turbine is acting is increased a certain amount. It appears probable that a diffuser of different proportions from those above indicated would operate with some advantage without being submerged. It is nearly always inconvenient to place the wheel entirely below low-water mark; up to this time, however, all that have been fitted up with a diffuser have been so placed; and indeed, to obtain the full effect of a fall of water, it appears essential, even when a diffuser is not used, that the wheel should be placed below the lowest level to which the water falls in the wheel-pit, when the wheel is in operation.

The action of the diffuser depends upon similar principles to that of diverging conical tubes, which, when of certain proportions, it is well known, increase the discharge; the author has not met with any experiments on tubes of this form discharging under water although there is good reason to believe that tubes of greater length and divergency would operate more effectively under water than when discharging freely in the air, and that results might be obtained that are now deemed impossible by most engineers.

Experiments on the same turbine, with and without a diffuser, show a gain in the *coefficient of effect*, due to the latter, of about three per cent. By the principles of living forces, and assuming that the motion of the water is free from irregularity, the gain should be about five per cent. The difference is due, in part at least, to the unstable equilibrium of water flowing through expanding apertures; this must interfere with the uniformity of the velocities of the fluid streams, at equal distances from the wheel.

Suspending the wheel on the top of the vertical shaft (Fig. A), *instead of running it on a step at the bottom.* This had been previously attempted, but not with such success as to warrant its general adoption. It has been accomplished with complete success by Mr. Boyden, whose mode is to cut the upper part of the shaft into a series of necks, and to rest the projecting parts upon corresponding parts of a box. A proper fit is secured by lining the box, which is of cast-iron, with Babbitt metal—a soft metallic composition consisting, principally, of tin; the cast-iron box is made with suitable projections and recesses, to support and retain the soft metal, which is melted and poured into it, the shaft being at the same time in its proper position in the box. It will readily be seen that a great amount of bearing-surface can be easily obtained by this mode, and also, what is of equal importance, it may be near the axis; the lining metal, being soft, yields a little if any part of the bearing should receive a great excess of weight. The cast-iron box is suspended on gimbals, similar to those usually adopted for mariners' compasses and chronometers, which arrangement permits the box to oscillate freely in all directions, horizontally, and prevents, in a great measure, all danger of breaking the shaft at the necks, in consequence of imperfections in the workmanship or in the adjustments. Several years' experience has shown that this arrangement, carefully constructed, is all that can be desired; and that a bearing thus constructed is as durable, and can be as readily oiled and taken care of, as any of the ordinary bearings in a manufactory.

The buckets are secured to the crowns of the wheel in a novel and much more perfect manner than had been previously used; the crowns are first turned to the required form, and made smooth; by ingenious machinery designed for the purpose, grooves are cut with great accuracy in the crowns, of the exact curvature of the buckets; mortices are cut through the crowns in several places in each groove; the buckets, or floats, are made with corresponding tenons, which project through the crowns, and are riveted on the bottom of the lower crown, and on the top of the upper crown; this construction gives the requisite strength and firmness, with buckets of much thinner iron than was necessary under any of the old arrangements; it also leaves the passages through the wheel entirely free from injurious obstructions.

In the year 1849, the manufacturing companies at Lowell purchased of Mr. Boyden the right to use all his improvements relating to turbines and other hydraulic motors. Since that time it has devolved upon the author, as the chief engineer of these companies, to design and superintend the construction of such turbines as might be wanted for their manufactories, and to aid him in this important undertaking, Mr. Boyden has communicated to him copies of many of his designs for turbines, together with the results of experiments upon a portion of them; he has communicated, however, but little theoretical information, and the author has been guided principally by a comparison of the most successful designs, and such light as he could obtain from writers on this most intricate subject.

Summary description of one of the turbines at the Tremont Mills, Lowell. Figs. A, B, C.

Fig. A is a vertical section of the turbine through the axis of the wheel shaft; Fig. B is a portion of the plan, on an enlarged scale, showing the disposition of the leading curves and buckets and diffuser; Fig. C is a cross section of the wheel and diffuser on an enlarged scale, and the more adjacent parts. The letters on the corresponding parts of the figures are the same.

The water is conveyed to the wheel of the turbine, from the forebay by a supply pipe, the greater portion of which, from the forebay downwards, is of wrought iron, and of gradually diminishing diameter towards the lower portion I, termed *the curbs*, which is of cast iron. The curbs are supported on columns, which rest on cast-iron supports firmly imbedded in the wheel-pit.

The *Disc* K, K′, K″, to which the guides for the water, or the leading curves, thirty-three in number, are attached, is suspended from the upper end of the cast-iron curb, by means of the *disc-pipes* M M.

The leading curves are of Russian iron, one-tenth of an inch in thickness. The upper corners of these, near the wheel, are connected by what is termed *the garniture* L, L′, L″, intended to diminish the contraction of the fluid vein when the regulating gate is fully raised.

The disc-pipe is very securely fastened, to sustain the pressure of the water on the disc. The escape of water, between the upper curb and the upper flange of the disc-pipe, is prevented by a band of leather on the outside, enclosed within an iron ring. This pipe is so fastened as to prevent its rotating in a direction opposite to that in which the water flows out.

The regulating gate is a cast-iron cylinder, R, enclosing the disc and curves, and which, raised or lowered by suitable machinery, regulates the amount of water let on the wheel B B′ B″, exterior to it.

The wheel consists of a central plate of cast-iron and two crowns, C,C, C′, C″, of the same material to which the buckets are attached. These pieces are all accurately turned, and polished, to offer the least obstruction in revolving rapidly in the water.

The buckets, made of Russian iron, are forty-four in number, and each $\frac{9}{64}$ of an inch thick. They are firmly fastened to the crowns.

The *vertical shaft* D, from which motion is communicated to the machinery by suitable gearing, is of wrought-iron. Instead of resting on a gudgeon, or step at bottom, it is suspended from a suspension box, E′, by which the collars at the top are enclosed. These collars are of steel, and are fastened to the upper portion of the shaft, which last can be detached from the lower portion.

The *suspension box* is lined with Babbitt metal, a soft composition consisting mostly of tin, and capable of sustaining a pressure of from 50 lbs. to 100 lbs. per square inch, without sensible diminution of durability. The box consists of two parts, for the convenience of fastening it on, or the reverse. The box rests upon the gimbal G, which is so arranged that the suspension box, the shaft, and the wheel can be lowered or raised, and the suspension box be allowed to oscillate laterally, so as to avoid subjecting it to any lateral strain.

The lower end of the shaft has a cast-steel pin, O, fixed to it. This is retained in its place by the step, which is made of three parts, and lined with case-hardened iron. The step can be adjusted by horizontal screws, by a small lateral motion given by them to it.

Rules for proportioning turbines. In making the designs for the Tremont and other turbines, the author has been guided by the following rules, which he has been led to, by a comparison of several turbines designed by Mr. Boyden, which have been carefully tested, and found to operate well.

Rule 1st. The sum of the shortest distances between the buckets should be equal to the diameter of the wheel.

Rule 2d. The height of the orifices of the circumference of the wheel should be equal to one-tenth of the diameter of the wheel.

Rule 3d. The width of the crowns should be four times the shortest distance between the buckets.

Rule 4th. The sum of the shortest distances between the curved guides, taken near the wheel, should be equal to the interior diameter of the wheel.

The turbines, from a comparison of which the above rules were derived, varied in diameter from twenty-eight inches to nearly one hundred inches, and operated on falls from thirty feet to thirteen feet. The author believes that they may be safely followed for all falls between five feet and forty feet, and for all diameters not less than two feet; and, with judicious arrangements in other respects, and careful workmanship, a useful effect of seventy-five per cent. of the power expended may be relied upon. For falls greater than forty feet, the second rule should be modified, by making the height of the orifices smaller in proportion to the diameter of the wheel.

Taking the foregoing rules as a basis, we may, by aid of the experiments on the Tremont turbine, establish the following formulas. Let

D = the diameter of the wheel at the outer extremities of the buckets.

d = the diameter at their inner extremities.

H = the height of the orifices of discharge, at the outer extremities of the buckets.

W = width of crowns of the buckets.

N = the number of buckets.

n = the number of guides.

P = the horse-power of the turbine, of 550 $^{\text{lbs. ft.}}$

h = the fall acting on the wheel.

Q = the quantity of water expended by the turbine, in cubic feet per second.

V = the velocity due the fall acting on the wheel.

V' = the velocity of the water passing the narrowest sections of the wheel.

v = the velocity of the interior circumference of the wheel; all velocities being in feet per second.

C = the coefficient of V', or the ratio of the real velocity of the water passing the narrowest sections of the wheel, to the theoretical velocity due the fall acting on the wheel.

The unit of length is the English foot.

It is assumed that the useful effect is seventy-five per cent. of the total power of the water expended.

According to Rule 1st, we have the sum of the widths of the orifices of discharge, equal to D. Then the sum of the areas of all the orifices of discharge is equal to $D\,H$.

By the fundamental law of hydraulics, we have

$$V = \sqrt{2\,g\,H}.$$

Therefore

$$V' = C\sqrt{2\,g\,H}.$$

For the quantity of water expended we have

$$Q = H\,D\,V' = H\,D\,C\sqrt{2\,g\,H}.$$

From the extremely interesting and accurate experiments of

Mr. Francis on the expenditure of water by one of the Tremont wheels, recorded in his work, the following data are obtained from it :—

For the sum of the widths of the orifices of discharge,

$$44 \times 0.18757 = 8.25308 \text{ feet.}$$

$Q = 138.1892$ cubic feet per second;

$h = 12.903$ feet;

$\sqrt{2g} = 8.0202$ feet.

Substituting these numerical results in the preceding value of Q, there obtains

$$138.1892 = 7.68692 \times 8.0202 \sqrt{12.903}\, C,$$

hence

$$C = 0.624.$$

By Rule 2d we have

$$H = 0.10\, D, \text{ hence } H\,D = 0.10\, D^2,$$

hence $Q = H\,D\,V' = 0.10\, D^2\, C \sqrt{2\,g\,h}$.

Calling the weight of a cubic foot of water 62.33 lbs., we have

$$P = \frac{0.75 \times 62.33}{550} Q\,h = 0.085\; Q\,h;$$

or, substituting for Q the value just found,

$$P = 0.0425\; D^2\, h \sqrt{h},$$

hence

$$D = 4.85 \sqrt{\frac{P}{h^{\frac{3}{2}}}}.$$

The number of buckets is to a certain extent arbitrary, and would usually be determined by practical considerations. Some of the ideas to be kept in mind are the following:

The pressure on each bucket is less, as the number is greater; the greater number will therefore permit of the use of thinner iron, which is important in order to obtain the best results. The width of the crowns will be less for a greater number of buckets. A narrow crown appears to be favorable to the useful effect, when the gate is only partially raised. As the spaces between the buckets must be proportionally narrower for a larger number of buckets, the liability to become choked up, either with anchor ice or other substances, is increased. The amount of power lost by the friction of the water against the surfaces of the buckets will not be materially changed, as the total amount of rubbing surface on the buckets will be nearly constant for the same diameter; there will be a little less on the crown, for the larger number. The cost of the wheel will probably increase with the number of buckets. The thickness and quality of the iron, or other metal intended to be used for the buckets, will sometimes be an element. In some water wrought iron is rapidly corroded.

The author is of opinion that a general rule cannot be given for the number of buckets; among the numerous turbines working rapidly in Lowell, there are examples in which the shortest distance between the buckets is as small as 0.75 of an inch, and in others as large as 2.75 inches.

As a guide in practice, to be controlled by particular circumstances, the following is proposed, to be limited to diameters of not less than two feet:—

$$N = 3\,(D + 10).$$

Taking the nearest whole number for the value of N.

The Tremont turbine is $8\frac{1}{3}$ in diameter, and, according to the proposed rule, should have fifty-five buckets instead of forty-four. With fifty-five buckets, the crowns should have a width of 7.2 inches instead of 9 inches. With the narrower width, it is probable that the useful effect, in proportion to the power expended, would have been a little greater when the gate was partially raised.

By the 3d rule, we have for the width of the crowns,

$$W = \frac{4\,D}{N};$$

and for the interior diameter of the wheel,

$$d = D - \frac{8\,D}{N}.$$

By the 4th rule, d is also equal to the sum of the shortest distances between the guides, where the water leaves them.

The number n of the guides is, to a certain extent, arbitrary. The practice at Lowell has been, usually, to have from a half to three-fourths of the number of the buckets; exactly half would probably be objectionable, as it would tend to produce pulsations or vibrations.

The proper velocity to be given to the wheel is an important consideration. Experiment 30 (the one above used for data) on the Tremont turbine gives the maximum coefficient of effect of that wheel; in that experiment, the velocity of the interior circumference of the wheel is 0.62645 of the velocity due to the fall acting on the wheel. By reference to other experiments, with the gate fully raised, it will be seen, however, that the coefficient of effect varies only about two per cent. from the maximum, for any velocity of the interior circumference, between fifty per cent. and seventy per cent. of that due to the fall acting upon the wheel. By reference to the experiments in which the gate is only partially raised, it will be seen that the maximum corresponds to slower velocities; and as turbines, to admit being regulated in velocity for variable work, must, almost necessarily, be used with a gate not fully raised, it would appear proper to give them a velocity such that they will give a good effect under these circumstances.

With this view, the following is extracted from the experiments in Table II.:—

Number of the experiment.	Height of the regulating gate in inches.	Ratio of the velocity of the interior circumference of the wheel, to the velocity due the fall acting upon the wheel, corresponding to the maximum coefficient of effect.
30	11.49	0.62645
62	8.55	0.56541
73	5.65	0.56205
84	2.875	0.48390

By this table it would appear that, as turbines are generally used, a velocity of the interior circumference of the wheel, of about fifty-six per cent. of that due to the fall acting upon the wheel, would be most suitable. By reference to the diagram at Plate VI,* it will be seen that at this velocity, when the gate is fully raised, the coefficient of effect will be within less than one per cent. of the maximum.

Other considerations, however, must usually be taken into account in determining the velocity; the most frequent is the variation of the fall under which the wheel is intended to operate. If, for instance, it were required to establish a turbine of a given power on a fall liable to be diminished to one-half by back-water, and that the turbine should be of a capacity to give the requisite power at all times, in this case the dimensions of the turbine must be determined for the smallest fall; but if it has assigned to it a velocity, to give the maximum effect at the smallest fall, it will evidently move too slow for the greatest fall, and this is the more objectionable, as, usually, when the fall is greatest the quantity of water is the least, and it is of the most importance to obtain a good effect. It would then be

* "Lowell Hydraulic Experiments."

usually the best arrangement to give the wheel a velocity corresponding to the maximum coefficient of effect, when the fall is greatest. To assign this velocity, we must find the proportional height of the gate when the fall is greatest; this may be determined approximately by aid of the experiments on the Tremont turbine.

We have seen that $P = 0.085\, Q h$.

Now, if h is increased to $2h$, the velocity, and consequently the quantity, of water discharged will be increased in the proportion of $\sqrt{h}$ to $\sqrt{2h}$; that is to say, the quantity for the fall $2h$ will be $\sqrt{2}\, Q$.

Calling P' the total power of the turbine on the double fall, we have

$$P' = 0.085 \sqrt{2}\, Q\, 2h,$$

or,

$$P' = 0.085 \times 2.8284\, Q h.$$

Thus, the total power of the turbine is increased 2.8284 times, by doubling the fall; on the double fall, therefore, in order to preserve the effective power uniform, the regulating gate must be shut down to a point that will give only $\frac{1}{2.8284}$ part of the total power of the turbine.

In Experiment 15, the fall acting upon the wheel was 12.888 feet, and the total useful effect of the turbine was 85625.3 lbs. raised one foot per second; $\frac{1}{2.8284}$ part of this is 30273.4 lbs.; consequently the same opening of gate that would give this last power on a fall of 12.888 feet, would give a power of 85625.3 lbs. raised one foot per second, on a fall of 2×12.888 feet $= 25.776$ feet. To find this opening of gate, we must have recourse to some of the other experiments.

In Experiment 73, the fall was 13.310 feet, the height of the gate 5.65 inches, and the useful effect 58830.1 lbs. In Experiment 83 the fall was 13.435 feet, the height of the gate

2.875 inches, and the useful effect 27310.9 lbs. Reducing both these useful effects to what they would have been if the fall was 12.888 feet,

the useful effect in experiment 73, $58830.1\left(\frac{12.888}{13.310}\right)^{\frac{3}{2}}=56054.5$;

" " " " " 83, $27310.9\left(\frac{12.888}{13.435}\right)^{\frac{3}{2}}=25660.1$.

By a comparison of the useful effects with the corresponding heights of gate, we find, by simple proportion of the differences, that a useful effect of 30273.4 lbs. raised one foot high per second, would be given when the height of the regulating gate was 3.296 inches.

By another mode:—

As 25660.1 : 2.875 :: 30273.4 : $2.875\times\frac{30273.4}{25660.1}=3.392$ in., a little consideration will show that the first mode must give too little, and the second too much; taking a mean of the two results, we have for the height of the gate, giving $\frac{1}{2.8284}$ of the total power of the turbine, 3.344 inches. Referring to Table II., we see that, with this height of gate, in order to obtain the best coefficient of useful effect, the velocity of the interior circumference of the wheel should be about one-half of that due to the fall acting upon the wheel; and by comparison of Experiments 74 and 84, it will be seen that, with this height of gate and with this velocity, the coefficient of useful effect must be near 0.50.

This example shows, in a strong light, the well-known defect of the turbine, viz., giving a diminished coefficient of useful effect at times when it is important to obtain the best results. One remedy for this defect would be, to have a spare turbine, to be used when the fall is greatly diminished; this arrangement would permit the principal turbine to be made nearly of the dimensions required for the greatest fall. As at other heights of the water economy of water is usually of less importance, the spare turbine might generally be of a cheaper construction.

To lay out the curve of the buckets, the author makes use of the following method:—

Referring to Fig. D, the number of buckets, N, having been determined by the preceding rules, set off the arc GI $= \frac{\pi D}{N}$.

Let $\omega =$ G H $=$ I P′, the shortest distance between the buckets; $t =$ the thickness of the metal forming the buckets.

Make the arc G K $= 5\,\omega$. Draw the radius O K, intersecting the interior circumference of the wheel at L; the point L will be the inner extremity of the bucket. Draw the directrix L M tangent to the inner circumference of the wheel. Draw the arc O N, with the radius $\omega + t$, from I as a centre; the other directrix, G P, must be found by trial, the required conditions being, that, when the line M L is revolved round to the position G T, the point M being constantly on the directrix G P, and another point at the distance M G $=$ R S, from the extremity of the line describing the bucket, being constantly on the directrix M L, the curve described shall just touch the arc N O. A convenient line for a first approximation may be drawn by making the angle O G P $= 11°$. After determining the directrix according to the preceding method, if the angle O G P should be greater than 12°, or less than 10°, the length of the arc G K should be changed to bring the angle within these limits.

The curve G S S′ S″ L, described as above, is nearly the quarter of an ellipse, and would be precisely so if the angle G M L was a right angle; the curve may be readily described, mechanically, with an apparatus similar to the elliptic trammel; there is, however, no difficulty in drawing it by a series of points, as is sufficiently obvious.

The trace adopted by the author for the corresponding guides is as follows:—

The number n having been determined, divide the circle in which the extremities of the guides are found into n equal parts V W, W X, etc.

Put ω' for the width between two adjoining guides,

and t' for the thickness of the metal forming the guides.

We have by Rule 4, $\omega' = \frac{d}{n}$.

With W as a centre, and the radius $\omega' + t'$, draw the arc Y Z; and with X as a centre, and the radius $2(\omega' + t')$, draw the arc A′ B′. Through V draw the portion of a circle, V C′, touching the arcs Y Z and A′ B′; this will be the curve for the essential portion of the guide. The remainder of the guide, C′ D′, should be drawn tangent to the curve C′ V; a convenient radius is one that would cause the curve C′ D′, if continued to pass through the centre O. This part of the guide might be dispensed with, except that it affords great support to the part C′ V, and thus permits the use of much thinner iron than would be necessary if the guide terminated at C′, or near it.

Collecting together the foregoing formulas for proportioning turbines, which, it is understood, are to be limited to falls not exceeding forty feet, and to diameters not less than two feet, we have for the horse power,

$$P = 0.0425 D^2 h \sqrt{h}\,;$$

for the diameter,

$$D = 4.85 \sqrt{\frac{P}{h\sqrt{h}}}\,;$$

for the quantity of water discharged per second,

$$Q = 0.5\, D^2 \sqrt{h}\,;$$

for the velocity of the interior circumference of the wheel, when the fall is not very variable,

$$v = 0.56 \sqrt{2 g h},$$

or,

$$v = 4.491 \sqrt{h};$$

for the height of the orifices of discharge,

$$H = 0.10 D;$$

for the number of buckets,

$$N = 3 (D + 10);$$

for the shortest distance between two adjacent buckets,

$$\omega = \frac{D}{N};$$

for the width of the crown occupied by the buckets,

$$W = \frac{4 D}{N};$$

for the interior diameter of the wheel,

$$d = D - \frac{8 D}{N};$$

for the number of guides,

$$n = 0.50 N \text{ to } 0.75 N;$$

for the shortest distance between two adjacent guides,

$$\omega' = \frac{d}{n}.$$

Table has been computed by these formulas.

For falls greater than forty feet, the height of the orifices in the circumference of the wheel should be diminished. The foregoing formulas may, however, still be made use of. Thus: supposing, for a high fall, it is determined to make the orifices three-fourths of that given by the formula; divide the given power, or quantity of water to be used, by 0.75, and use the quotient in place of the true power or quantity, in determining the dimensions of the turbine. No modifications of the dimensions will be necessary, except that $\frac{1}{10}$ of the diameter of the turbine should be diminished to $\frac{3}{40}$ of the diameter, to give the height of the orifices in the circumference.

It is plain, from the method by which the preceding formulas have been obtained, that they cannot be considered as established, but should only be taken as guides in practical applications, until some more satisfactory are proposed, or the intricacies of the turbine have been more fully unravelled. The turbine has been an object of deep interest to many learned mathematicians, but up to this time the results of their investigations, so far as they have been published, have afforded but little aid to hydraulic engineers.

Diffuser.—As previously stated, the principles involved in the flow of water through a diverging tube find a useful application in Mr. Boyden's diffuser. This invention, applied to a turbine water wheel 104.25 inches in diameter, and about seven hundred horse-power, is represented on Fig. B at X′ and on Fig. C at X″.

The diffuser is supported on iron pillars from below. The wheel is placed sufficiently low to permit the diffuser to be submerged at all times when the wheel is in operation, that being essential to the most advantageous operation of the diffuser.

When the speed gate is fully raised, the wheel moves with the velocity which gives its greatest coefficient of useful effect. On leaving the wheel it necessarily has considerable velocity, which would involve a corresponding loss of power, except for the effect of the diffuser, which utilizes a portion of it. When operating under a fall of thirty-three feet, and the speed gate is raised to its full height, this wheel discharges about 219 cubic feet per second. The area of the annular space where the water enters the diffuser, is $0.802 \times 8.792\,\pi = 22.152$ square feet; and if the stream passes through this section radially, its mean velocity must be $\frac{219}{22.152} = 9.886$ feet per second, which is due to a head of 1.519 feet. The area of the annular space where the water leaves the diffuser is $1.5 \times 15.333\ \pi = 72.255$ square feet, and the mean velocity $\frac{219}{72.255} = 3.031$ feet per second, which is due to a head of 0.143 feet. According to this the saving of head due to the diffuser is $1.519 - 0.143 = 1.376$ feet, being $\frac{1.376}{33 - 1.519}$ or about $4\frac{3}{8}$ per cent. of the head available without the diffuser, which is equivalent to a gain in the coefficient of useful effect to the same extent. Experiments on the same turbine, with and without the diffuser, have shown a gain due to the latter of about three per cent. in the coefficient of useful effect. The diffuser adds to the coefficient of useful effect by increasing the velocity of the water passing through the wheel, and it must of course increase the quantity of water discharged in the same proportion. If it increases the available head three per cent., the velocity, which varies as the square root of the head, must be increased in the same proportion. The power of the wheel, which varies as the product of the head into the quantity of water discharged, must be increased about 4.5 per cent.

EXPLANATION OF FIGURES.

Fig. A. Section through the axis of the Turbine without the Diffuser.

I Cast-iron Curbs through which the water passes from the wrought-iron supply-pipe to the Disc.

K Cast-iron Disc on which the Guide Curves are fastened.

L Garniture fitted to lower end of Lower Curb.

M Disc-pipe suspending the Disc from Upper Curb.

N Columns of cast-iron supporting Curbs.

R Regulating Gate of cast-iron.

S Brackets for raising and lowering the Gate.

B Wheel.

C, C Crowns of the Wheel between which the curved buckets are inserted.

D Main Shaft of the Wheel.

E′ Suspension Box, lined with babbit metal, from which the Wheel hangs by the cast-steel Collars around the upper end of the Shaft.

F Upper portion of Shaft fastened to lower portion, with bearings F′ of cast-iron lined with babbit metal.

G Gimbal on which the Suspension Box E′ rests.

H Support of the Gimbal.

O Step to receive cast-steel Pin on lower end of Shaft.

Fig. A. Plan of Disc, K″, Garniture, L′, Wheel, C′, and Diffuser, M′ N′.

Fig. B. Section of Wheel, C″,Garniture, L″, Regulating Gate, R″, and upper and lower Crowns of the Diffuser.

Fig. D. Diagram for laying out the curves of the Buckets and Guide Curves.

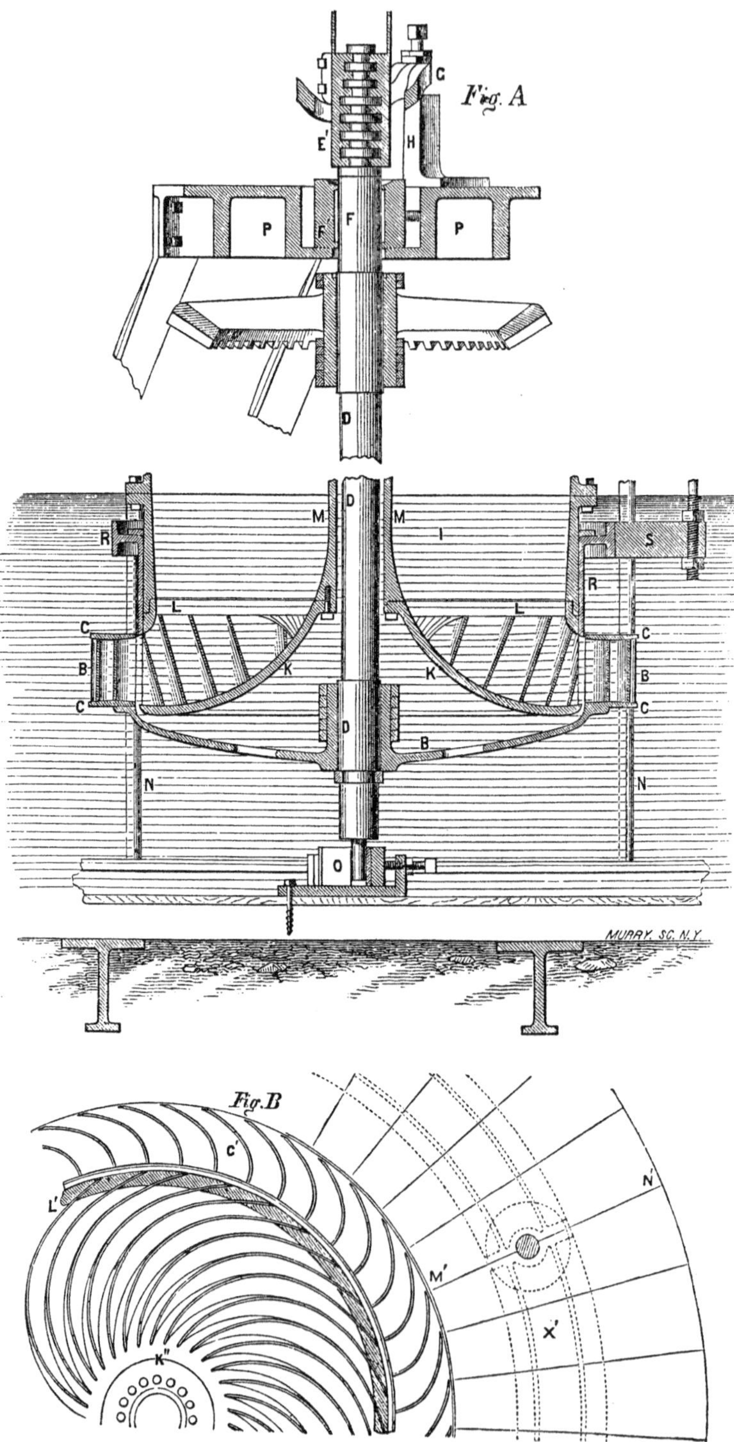
Fig. A
C
E'
H
P
F'
F
P
D
D
M
M
I
R
S
R
L
L
C
C
B
B
K
K
C
C
D
B
N
N
O
MURRY. SC. N.Y.
Fig. B
C'
L'
N'
M'
X'
K"

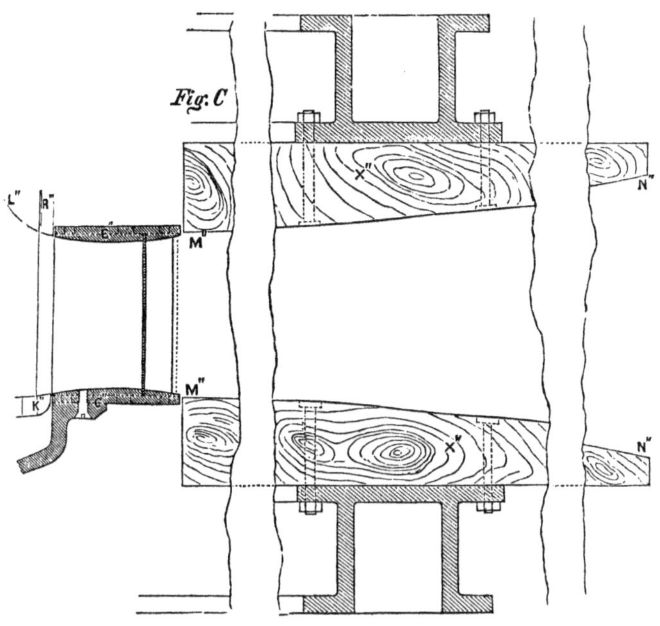

Fig. C

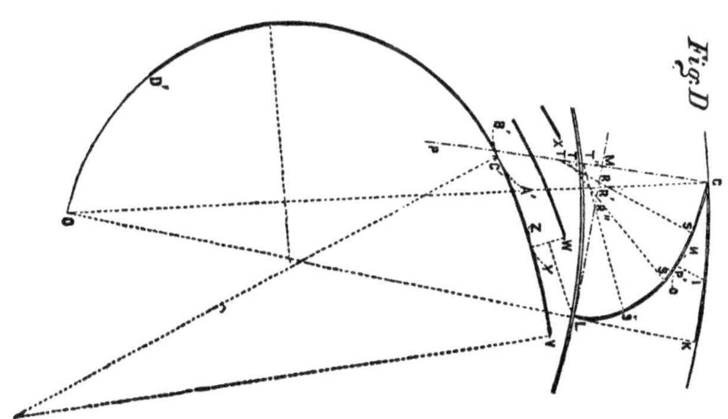

Fig. D

www.ingramcontent.com/pod-product-compliance
Lightning Source LLC
LaVergne TN
LVHW021401110826
845150LV00007B/1736

* 9 7 8 1 4 2 5 5 1 3 0 9 2 *